How to . . .

get the most from your
COLES NOTES

Key

Basic co

Close Up

Additional hints, notes, tips
or background information.

Watch Out!

Areas where problems
frequently occur.

Quick Tip

Concise ideas to help you
learn what you need to know.

Remember This!

Essential material for
mastery of the topic.

How to get an *A* in . . .

Calculus

Limits & derivatives

Integration

Problems & sample

exams

COLES NOTES have been an indispensable aid to students on five continents since 1948.

COLES NOTES now offer titles on a wide range of general interest topics as well as traditional academic subject areas and individual literary works. All COLES NOTES are written by experts in their fields and reviewed for accuracy by independent authorities and the Coles Editorial Board.

COLES NOTES provide clear, concise explanations of their subject areas. Proper use of COLES NOTES will result in a broader understanding of the topic being studied. For academic subjects, COLES NOTES are an invaluable aid for study, review and exam preparation. For literary works, COLES NOTES provide interesting interpretations and evaluations which supplement the text but are not intended as a substitute for reading the text itself. Use of the NOTES will serve not only to clarify the material being studied, but should enhance the reader's enjoyment of the topic.

© Copyright 2001 and Published by
COLES PUBLISHING. A division of Prospero Books
Toronto – Canada
Printed in Canada

Cataloguing in Publication Data
Erdman, Wayne 1954–

How to get an A in ... Calculus

(Coles notes) ISBN 0-7740-0597-4

1. Calculus – Problems, exercises, etc.   2. Calculus.
I. Title.   II. Series

QA303.E72 1999   515'.1   C99-930425-9

Publisher: Nigel Berrisford
Editing: Paul Kropp Communications
Book design and layout: Karen Petherick, Markham, Ontario

Manufactured by Webcom Limited
Cover finish: Webcom's Exclusive DURACOAT

Contents

Introduction

Calculus can be described as the study of changing quantities. Change can be seen in the velocity of a car, the growth of a city, the slope of a graph, marginal costs of goods and fluctuations in temperatures. Calculus is used in physics, medicine, engineering, economics, business and any profession where change can be modelled with equations. It is also a unifying link between other areas of mathematics, such as algebra, trigonometry, geometry and statistics. This is why Calculus, as a math course, is so valuable in the sciences, social sciences and business studies at university and college.

Because the study of Calculus is a great departure from traditional math courses found in high school, students tend to struggle with its complexity. Problem solving is combined with graphing, trigonometry, geometry and algebra, along with new concepts of limits, derivatives and integration. It is a rare student who flies through a Calculus course without some difficulties. *How to ... Get an A in Calculus* provides explanations, techniques for developing model solutions, memory hints and study techniques, all tailored to helping you through your Calculus course. If you are a senior high school or first year university or college student, this book will be a valuable companion to your studies.

Other Coles Notes covering topics in senior mathematics:

Topics in senior math are frequently interconnected at each grade level. These additional titles from Coles Notes will help you master them all:

How to Get an A in ...

- Permutations, Combinations and Probability
- Senior Algebra
- Sequences and Series
- Statistics and Data Analysis
- Trigonometry and Circle Geometry

Limits

INTRODUCTION TO LIMITS

Calculus is the exploration of change ... change in the form of slope ... change in the form of velocity ... change in the form of area.

Fundamental to these explorations of change is the concept of a limit.

First, an analogy.

Consider a spaceship invading a distant planet. There is a force-field around the planet. The spaceship can get very close to the force-field, but cannot pass through it.

In Calculus, this force-field would be called a **limit.**

> A **limit** is the value that a function **approaches**
> (without necessarily being equal to)
> as x approaches a specific value

3

Now, let's consider the function $y=2x^2 - 3$ near the point (2,5), hilighted as shown. Treat the point (2,5) as the force field in the analogy above.

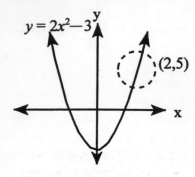

The value of 2 for x can be approached from the left (values less than 2) or from the right (values greater than 2).

x	y
1.5	1.5
1.9	4.22
1.99	4.92
1.999	4.992

x	y
2.5	9.5
2.1	5.82
2.01	5.08
2.001	5.008

As x approaches 2 from the left, we can see that y approaches 5. As x approaches 2 from the right, again we can see that y approaches 5. Because the **solution is the same from both sides**, we can say that the **limit exists** and is equal to 5. In limit notation, this is written as:

$$\lim_{x \to 2}\left(2x^3 - 3\right) = 5$$

Remembering that $y=2x^2 - 3$ is a polynomial-type function, we can say that, because all polynomials are continuous (i.e.,they have no gaps in their graphs), their limits can be evaluated by substitution:

For any polynomial function f(x), the limit can be evaluated by substitution:

$$\lim_{x \to 0} f(x) = f(a)$$

4

Example 1

Evaluate each of the following limits.

All can be evaluated by substitution, because they are polynomial-type functions

(a) $\lim\limits_{x \to 0}\left(2 - 3x^2\right)$ $= 2 - 3(0)$

$\qquad\qquad = 2$

(b) $\lim\limits_{x \to -1}\left(1 - 2x + x^4\right)$ $= 1 - 2(-1) + (-1)^4$ *Substituting shows that we can simply add the limits of the individual terms*

$\qquad\qquad\qquad = 4$

(c) $\lim\limits_{x \to 2}\left(2x - 5\right)\left(x^2 + 6\right)$ $= \left[2(2) - 5\right]\left[(2)^2 + 6\right]$ *Substituting shows that we can simply multiply the limits of the individual factors.*

$\qquad\qquad\qquad = (-1)(10)$

$\qquad\qquad\qquad = -10$

Based on the above examples, the following properties can be stated:

The limit of a **sum** is equal to
the sum of the limits:

$$\lim\limits_{x \to a}\left[f(x) + g(x)\right] = \lim\limits_{x \to a} f(x) + \lim\limits_{x \to a} g(x)$$

The limit of a **product** is equal to
the product of the limits:

$$\lim\limits_{x \to a} f(x)g(x) = \lim\limits_{x \to a} f(x) \times \lim\limits_{x \to a} g(x)$$

Example 2

This example is of a rational variable expression, or a polynomial divided by another polynomial. As long as the denominator is not zero (indicating undefined), we can simply substitute for x and simplify.

$$\lim_{x \to 0}\left(\frac{x^2 - 3x + 2}{x + 2}\right) = \frac{0^2 - 3(0) + 2}{0 + 2}$$

$$= 1$$

Example 3

This example contains non-integer exponents which, in most cases can be evaluated by substitution.

$$\lim_{x \to 9}\left(x^{\frac{1}{2}} - 3\right) = \sqrt{9} - 3$$

$$= 0$$

INDETERMINATE FORMS: 0/0

The following examples provide what is called **indeterminate form** because, after substituting, the result is $\frac{0}{0}$. This expression is meaningless. Fortunately, when $\frac{0}{0}$ occurs, the original expression can usually be simplified in some way. These will be discussed in Examples 4 through 9.

Example 4

$$\lim_{x \to 1}\frac{(x - 1)(x + 3)}{x - 1} = \frac{0}{0}$$

Because of the indeterminate form, we need to develop a different method of evaluating this limit. Notice that the problem is caused by the factor of x–1 in both the numerator and denominator. Thus it would help to divide through by x–1.

$$\lim_{x \to 1}\frac{(x - 1)(x + 3)}{x - 1} = \lim_{x \to 1}(x + 3)$$

$$= 1 + 3$$

$$= 4$$

6

Example 5

$$\lim_{x \to -3} \frac{x^2 + x - 6}{x + 3} = \lim_{x \to -3} \frac{(x+3)(x-2)}{x+3}$$
$$= \lim_{x \to -3} (x - 2)$$
$$= -3 - 2$$
$$= -5$$

FACTORING

In examples such as this one, try factoring first.

Example 6

$$\lim_{x \to 2} \frac{\frac{1}{x} - \frac{1}{2}}{x - 2} = \lim_{x \to 2} \frac{\frac{2}{2x} - \frac{x}{2x}}{x - 2}$$
$$= \lim_{x \to 2} \frac{2 - x}{2x(x - 2)}$$
$$= \lim_{x \to 2} \frac{-1}{2x}$$
$$= \frac{-1}{2(2)}$$
$$= -\frac{1}{4}$$

Find the common denominator of the fractions in the numerator.

Then combine the denominators before simplifying.

Example 7

$$\lim_{x \to 3} \frac{x^3 - 27}{x - 3} = \lim_{x \to 3} \frac{(x-3)(x^2+3x+9)}{x-3}$$
$$= \lim_{x \to 3} (x^2 + 3x + 9)$$
$$= 3^2 + 3(3) + 9$$
$$= 27$$

Recall factoring a difference of cubes:

$$x^3 - a^3 = (x - a)(x^2 + ax + a^2)$$

Example 8

$$\lim_{x \to 4} \frac{\sqrt{x} - 2}{x - 4} = \frac{\sqrt{4} - 2}{4 - 4} = \frac{0}{0}$$

Factoring would be possible here, but an easier method would be to multiply the numerator and denominator by the conjugate of $\sqrt{x} - 2$.

7

$$\lim_{x \to 4} \frac{\sqrt{x}-2}{x-4} \times \frac{\sqrt{x}+2}{\sqrt{x}+2} = \lim_{x \to 4} \frac{x-4}{(x-4)(\sqrt{x}+2)}$$

$$= \lim_{x \to 4} \frac{1}{\sqrt{x}+2}$$

$$= \frac{1}{\sqrt{4}+2}$$

$$= \frac{1}{4}$$

Example 9

Evaluate $\lim_{x \to 0} \dfrac{\sqrt{16+x}-4}{x}$

$$= \lim_{x \to 0} \frac{\sqrt{16+x}-4}{x} \times \frac{\sqrt{16+x}+4}{\sqrt{16+x}+4}$$

$$= \lim_{x \to 0} \frac{(16+x)-16}{x(\sqrt{16+x}+4)}$$

$$= \lim_{x \to 0} \frac{x}{x(\sqrt{16+x}+4)}$$

$$= \lim_{x \to 0} \frac{1}{\sqrt{16+x}+4}$$

$$= \frac{1}{8}$$

What looks like a very difficult problem becomes a relatively easy one when using the conjugate.

LIMITS FROM THE LEFT AND RIGHT SIDES

As we discussed in the introduction to this chapter, a limit only exists if the limits approaching from the left and the right both exist and are equal.

Remember the graph of $y = 2x^2 - 3$.
We discussed the limits from either side.

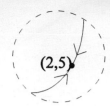

To indicate a limit approaching from the left, we use a superscript "-" sign:

$$\lim_{x \to 2^-}\left(2x^2 - 3\right) = 5$$

To indicate a limit approaching from the right, we use a superscript "+" sign:

$$\lim_{x \to 2^+}\left(2x^2 - 3\right) = 5$$

Both of these limits exist and are equal.

$$\therefore \lim_{x \to 2}\left(2x^2 - 3\right) = 5$$

As we saw earlier, existence and equality of the limits are not a problem with most functions. The following examples illustrate some exceptions.

Example 10

Evaluate the following limit, if it exists.

$$\lim_{x \to 0}\sqrt{x}$$

Evaluate the limits from the right and left sides first.

$$\lim_{x \to 0^+}\sqrt{x} = 0$$

$\lim_{x \to 0^-}\sqrt{x}$ *does not exist because* $x < 0$

      and the square root of a negative

      number is undefined.

$\therefore \lim_{x \to 0}\sqrt{x}$ *does not exist.*

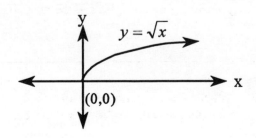

Example 11

Evaluate $\lim_{x \to 8} \sqrt{8-x}$

$\lim_{x \to 8^+} \sqrt{8-x}$ *gives square roots of negative numbers when substituting because 8⁺ means numbers slightly greater than 8 and 8-x would be negative*

$\therefore \lim_{x \to 8} \sqrt{8-x}$ *does not exist.*

Example 12

Evaluate $\lim_{x \to 7} |x - 7|$

$\lim_{x \to 7^-} |x - 7| = |7^- - 7|$

        $= |0^-|$

        $= 0$

$\lim_{x \to 7^+} |x - 7| = |7^+ - 7|$

        $= |0^+|$

        $= 0$

$\therefore \lim_{x \to 7} |x - 7| = 0$

With absolute values, we can see that, (after all of the calculations are done inside the absolute value brackets), the results are the same.

Example 13

Evaluate $\lim_{x \to 0} f(x)$

where $f(x) = \begin{cases} x - 1 & \text{if } x < 0 \\ 2 & \text{if } x = 0 \\ x^2 - 1 & \text{if } x > 0 \end{cases}$

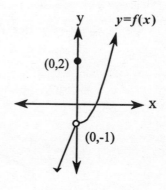

$\lim_{x \to 0^-} f(x) = 0 - 1 = -1$

$\lim_{x \to 0^+} f(x) = 0^2 - 1 = -1$

$\therefore \lim_{x \to 0} f(x) = -1$

*Even though f(0)=2, because the function **approaches** −1 from **both** sides, the limit is −1.*

*Because of the mathematical "hole" at x=0, the function f(x) is considered to be **discontinuous**.*

Example 14

The Big Eater Buffet offers one price meals; the price is based on the customer's age. Their prices are shown in the table:

Age (years)	under 2	under 8	under 15	15 or older
Price in $	2.00	5.00	10.00	15.00

(a)   Draw the graph of the price, y, of eating at the Big Eater Buffet as a function of the customer's age, x.

(b)   Where is the graph discontinuous?

(a)

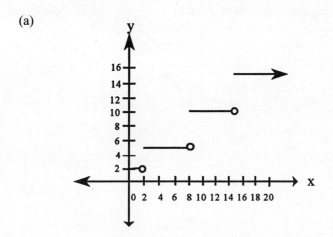

(b) The graph is discontinuous at $x = 2$, $x = 8$ and $x = 15$.

LIMITS INVOLVING INFINITY

People generally think of infinity in the context of time. In mathematics, however, infinity refers to numbers that grow without bound.

Example 15

Examine the graph of the function $y = \dfrac{1}{x}$

and investigate the limit $\lim\limits_{x \to 0} \dfrac{1}{x}$

From the graph, we can see that, as x approaches 0 from the left, the curve approaches negative infinity. From the right, the graph approaches positive infinity.

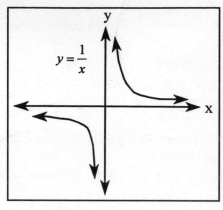

$\lim\limits_{x \to 0^-} \dfrac{1}{x} = -\infty$ *(Does not exist)*

$\lim\limits_{x \to 0^+} \dfrac{1}{x} = +\infty$ *(Does not exist)*

$\therefore \lim\limits_{x \to 0} \dfrac{1}{x}$ *does not exist.*

Using tables of value we can see how 1/x grows without bound:

x	$1/x$
-0.5	-2
-0.1	-10
-0.001	-1000
-0.000001	-1000000

x	$1/x$
0.5	2
0.1	10
0.001	1000
0.000001	1000000

Therefore the limit does not exist and the function is discontinuous at x=0.

Example 16

Evaluate $\lim\limits_{x \to 5} \dfrac{x^2+3}{x-5}$

If we substitute $x = 5$: $\lim\limits_{x \to 5} \dfrac{x^2+3}{x-5} = \dfrac{28}{0}$

$$= \pm\infty$$

$$\therefore \lim\limits_{x \to 5} \dfrac{x^2+3}{x-5} \quad \textit{does not exist.}$$

Example 17

Using the graph in Example 14, evaluate the limits $\lim\limits_{x \to \infty} \dfrac{1}{x}$ and $\lim\limits_{x \to \infty} \dfrac{1}{x}$.

From the graph, we can see that, as x approaches ∞, the function approaches 0.

We can also see that, as x approaches −∞, the function approaches 0.

$$\therefore \lim\limits_{x \to -\infty} \dfrac{1}{x} = 0$$

$$\textit{and} \quad \lim\limits_{x \to \infty} \dfrac{1}{x} = 0$$

 It is important to remember that this technique works only for $\frac{a}{x}$, where $a \neq 0$.

Example 18

Evaluate $\lim\limits_{x \to \infty}\left(2x^2 - 3x + 4\right)$

When we substitute ∞ for x: $\lim\limits_{x \to \infty}\left(2x^2 - 3x + 4\right) = \infty$

$$\therefore \textit{ it does not exist.}$$

Example 19

Evaluate $\lim\limits_{x \to \infty} \dfrac{5x^2 + 3x + 1}{2x^2 - 6x - 8}$

13

When we substitute as in Example 17, we get $\frac{\infty}{\infty}$ which is an indeterminate form.

*Using the methods of Examples 16 and 17, try multiplying each term in the numerator and denominator by $\frac{1}{x^2}$ (which contains the variable power with the **greatest exponent**).*

$$\lim_{x\to\infty}\frac{5x^2+3x+1}{2x^2-6x-8}\times\frac{\dfrac{1}{x^2}}{\dfrac{1}{x^2}}=\lim_{x\to\infty}\frac{\dfrac{5x^2}{x^2}+\dfrac{3x}{x^2}+\dfrac{1}{x^2}}{\dfrac{2x^2}{x^2}-\dfrac{6x}{x^2}-\dfrac{8}{x^2}}$$

By multiplying by $\frac{1}{x^2}$ we are able to substitute $x=\infty$ and use the technique from example 17.

$$=\lim_{x\to\infty}\frac{5+\dfrac{3}{x}+\dfrac{1}{x^2}}{2-\dfrac{6}{x}-\dfrac{8}{x^2}}$$

$$=\frac{5+0+0}{2-0-0}$$

$$=\frac{5}{2}$$

Summary of limits

For a limit to exist, the limits from the left and right sides must both exist and be equal.

POLYNOMIAL FUNCTIONS:
Substitute for x

INDETERMINATE FORMS, 0/0:
(a) Factor (if possible) and reduce
(b) If there is a square root, try multiplying the numerator and denominator by the conjugate

DENOMINATOR ZERO:
If the denominator (but not the numerator) approaches zero, the limit approaches infinity and is therefore undefined.

DENOMINATOR INFINITY:
If the denominator (but not the numerator) approaches infinity, the limit approaches zero.

INDETERMINATE FORM, ∞/∞ :
(a) Try similar methods as for 0/0
(b) Divide through by the variable power with the greatest exponent.

CONTINUITY:

If $\lim\limits_{x \to a} f(x) \neq f(a)$ then the function is discontinuous.

PRACTICE EXERCISE 1

1. Evaluate each of the following limits, if it exists.

   (a) $\lim\limits_{x \to 3} (2x + 7)$

   (b) $\lim\limits_{x \to 4} \dfrac{x - 4}{x + 2}$

   (c) $\lim\limits_{x \to 2} \dfrac{x + 3}{x - 3}$

   (d) $\lim\limits_{x \to 9} \sqrt{x + 7}$

   (e) $\lim\limits_{x \to 0} \dfrac{x^2 - 3x + 5}{x^2 + 5x + 5}$

   (f) $\lim\limits_{x \to -4} \sqrt{x^2 - 4x + 12}$

   (g) $\lim\limits_{x \to 0} \dfrac{\sqrt{x + 5} - \sqrt{5}}{x}$

   (h) $\lim\limits_{x \to -2} \dfrac{x + 2}{x^2 - 4}$

   (i) $\lim\limits_{x \to 3} \dfrac{x^2 - 2x - 3}{x^2 - 4x + 3}$

   (j) $\lim\limits_{x \to 3} \dfrac{x^3 - 27}{x^2 - 9}$

   (k) $\lim\limits_{x \to 25} \dfrac{x - 25}{\sqrt{x} - 5}$

   (l) $\lim\limits_{x \to 4} \dfrac{\frac{1}{x} - \frac{1}{4}}{x - 4}$

   (m) $\lim\limits_{x \to 6} \dfrac{1}{(x - 6)^2}$

   (n) $\lim\limits_{x \to 2} \dfrac{x + 3}{x - 2}$

   (o) $\lim\limits_{x \to \infty} \dfrac{x^4 - 3x^2 + 7x + 2}{3x^3 + 4x^2 - 5x - 6}$

   (p) $\lim\limits_{x \to \infty} \left(\dfrac{3}{4}\right)^x$

2. Evaluate each of the following limits, if it exists.

   (a) $\lim\limits_{x \to 5} \dfrac{|x - 5|}{x - 5}$

   (b) $\lim\limits_{x \to 2} f(x)$ for $f(x) = \begin{cases} 4x - 1 & \text{if } x < 2 \\ 2x & \text{if } x = 2 \\ x^2 & \text{if } x > 2 \end{cases}$

   Indicate where the graph of $f(x)$ would be discontinuous

   (c) $\lim\limits_{x \to 6} \sqrt{x - 6}$

   (d) $\lim\limits_{x \to 3} \dfrac{x}{x^2 - 9}$

   (e) $\lim\limits_{x \to \infty} \dfrac{2^x}{x^2}$ (hint: use a table of values)

CHAPTER TWO

Introduction to derivatives

SLOPES OF SECANTS AND TANGENTS

Shown below is the graph of $y=x^2$, with a secant through the points $(-2,4)$ and $(3,9)$.

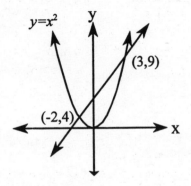

Remember that a secant is a line passing through 2 points on a curve. To find the slope of the secant, we use the standard slope formula in two different, but equivalent, ways:

$$Slope = \frac{\Delta y}{\Delta x}$$

$$= \frac{9-4}{3-(-2)}$$

$$= 1$$

$$Slope = \frac{f(x+\Delta x)-f(x)}{\Delta x} \quad \text{where } \Delta x = 3-(-2)=5$$

$$= \frac{f(-2+5)-f(-2)}{3-(-2)}$$

$$= \frac{9-4}{3-(-2)}$$

$$= 1$$

The importance of calculus enters when we try to find the slope of a tangent. Consider the same curve, $y=x^2$, with a tangent at the point (2,4):

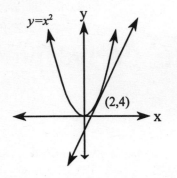

The difficulty here is that we have only a single point of contact. We could try to find a second point elsewhere on the tangent, but this would only give us an approximate value for the slope.

In the graph below, we have the same curve of $y = x^2$, with secants S1, S2, and S3 moving around the graph to become a tangent, T, at the point (2,4).

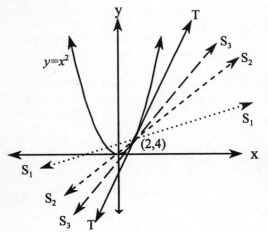

As the secants shift around the curve towards the tangent, we can see that Δx gets very small and approaches 0. Therefore, we can determine the slope of the tangent by taking **the limit of the slope of the secant as** $\Delta x \to 0.$

$$Slope = \lim_{\Delta x \to 0} \frac{\Delta y}{\Delta x}$$

$$= \lim_{\Delta x \to 0} \frac{f(x + \Delta x) - f(x)}{\Delta x}$$

$$= \lim_{\Delta x \to 0} \frac{(x + \Delta x)^2 - x^2}{\Delta x}$$

$$= \lim_{\Delta x \to 0} \frac{x^2 + 2x(\Delta x) + (\Delta x)^2 - x^2}{\Delta x}$$

$$= \lim_{\Delta x \to 0} \frac{2x(\Delta x) + (\Delta x)^2}{\Delta x}$$

$$= \lim_{\Delta x \to 0} (2x + \Delta x)$$

$$= 2x$$

Because f(x) represents the function x², we substitute x+Δx in x².

At x = 2, slope = 2(2)
           = 4

Slope of the tangent to $y=f(x)$ at (x_1, y_1) can be found through what are known as first principles.

$$\lim_{\Delta x \to 0} \frac{f(x_1 + \Delta x) - f(x_1)}{\Delta x}$$

Example 1

(a) Find the slope and an equation of the tangent to $y = \sqrt{x+1}$ at the point (3,2).

(b) Graph both the curve and the tangent.

$$Slope = \lim_{\Delta x \to 0} \frac{f(x + \Delta x) - f(x)}{\Delta x}$$

$$= \lim_{\Delta x \to 0} \frac{\sqrt{x + \Delta x + 1} - \sqrt{x + 1}}{\Delta x}$$

$$= \lim_{\Delta x \to 0} \frac{\sqrt{x + \Delta x + 1} - \sqrt{x + 1}}{\Delta x} \times \frac{\sqrt{x + \Delta x + 1} + \sqrt{x + 1}}{\sqrt{x + \Delta x + 1} + \sqrt{x + 1}}$$

Multiply by the conjugate

$$= \lim_{\Delta x \to 0} \frac{x + \Delta x + 1 - (x + 1)}{\Delta x(\sqrt{x + \Delta x + 1} + \sqrt{x + 1})}$$

$$= \lim_{\Delta x \to 0} \frac{\Delta x}{\Delta x(\sqrt{x + \Delta x + 1} + \sqrt{x + 1})}$$

$$= \lim_{\Delta x \to 0} \frac{1}{(\sqrt{x + \Delta x + 1} + \sqrt{x + 1})}$$

$$= \frac{1}{2\sqrt{x + 1}}$$

At $x = 3$, *Slope* $= \dfrac{1}{2\sqrt{3 + 1}}$

$$= \frac{1}{4}$$

To find the equation, use

$$\frac{y - y_1}{x - x_1} = slope$$

$$\frac{y - 2}{x - 3} = \frac{1}{4}$$

$$4y - x = 5$$

$y = \sqrt{x + 1}$

(3,2)

(-1,0)

Example 2

Determine the equation of the tangent to $y = \dfrac{1}{x}$ at the point (1,1). Graph the curve and tangent.

Using first principles,

$$Slope = \lim_{\Delta x \to 0} \frac{\dfrac{1}{x + \Delta x} - \dfrac{1}{x}}{\Delta x}$$

$$= \lim_{\Delta x \to 0} \frac{\dfrac{x - (x + \Delta x)}{x(x + \Delta x)}}{\Delta x}$$

$$= \lim_{\Delta x \to 0} \frac{-\Delta x}{\Delta x \, x(x + \Delta x)}$$

$$= \lim_{\Delta x \to 0} \frac{-1}{x(x + \Delta x)}$$

$$= \frac{-1}{x^2}$$

$$Slope \Big|_{x=1} = \frac{-1}{1^2}$$

$$= -1$$

Tangent equation:

$$\frac{y - 1}{x - 1} = -1$$

$$y = -x + 2$$

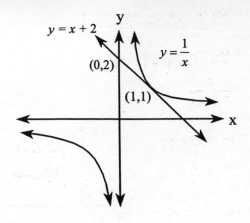

DERIVATIVES BY FIRST PRINCIPLES

The examples so far have been exercises in finding what is known as the **derivative** of the function. The derivative is the equation found after taking the limit using first principles (see the box preceding Example 1).

Symbols commonly used to identify a derivative are $D_x y$, $\dfrac{dy}{dx}$, $f'(x)$ and y'.

They are read as, "the derivative of y with respect to x" or "f prime of x"

The symbol most commonly used in textbooks is $\dfrac{dy}{dx}$ and that's what we will use here.

EXAMPLE 3

Determine the derivative of $y = x^3 - 2$ by first principles.

$$\frac{dy}{dx} = \lim_{\Delta x \to 0} \frac{\left[(x + \Delta x)^3 - 2\right] - (x^3 - 2)}{\Delta x}$$

$$= \lim_{\Delta x \to 0} \frac{[x^3 + 3x^2 \Delta x + 3x(\Delta x)^2 + (\Delta x)^3 - 2] - x^3 + 2}{\Delta x}$$

$$= \lim_{\Delta x \to 0} \frac{3x^2 \Delta x + 3x(\Delta x)^2 + (\Delta x)^3}{\Delta x}$$

$$= \lim_{\Delta x \to 0}\left[3x^2 + 3x\Delta x + \left(\Delta x^2\right)\right]$$
$$= 3x^2$$

Example 4

Determine the derivative of the function $y = \dfrac{2}{x-1}$ using first principles.

$$\frac{dy}{dx} = \lim_{\Delta x \to 0} \frac{\dfrac{2}{(x+\Delta x)-1} - \dfrac{2}{x-1}}{\Delta x}$$

$$= \lim_{\Delta x \to 0} \frac{2(x-1)-2(x+\Delta x -1)}{\Delta x(x+\Delta x -1)(x-1)}$$

$$= \lim_{\Delta x \to 0} \frac{-2\Delta x}{\Delta x(x+\Delta x -1)(x-1)}$$

$$= \lim_{\Delta x \to 0} \frac{-2}{(x+\Delta x -1)(x-1)}$$

$$= \frac{-2}{(x-1)^2}$$

VELOCITY PROBLEMS

Velocity is the change in distance from a fixed point over a given interval of time.

Example 5 AVERAGE VELOCITY

The displacement (distance from a fixed point), in metres, of an object is given by the equation $s = t^2 + 5$, where t is the time in seconds.

Calculate the average velocity between 2 seconds and 5 seconds.

$$At \ \ t = 2, \ \ s = 2^2 + 5$$
$$= 9$$
$$At \ \ t = 5, \ \ s = 5^2 + 5$$
$$= 30$$
$$Average \ \ velocity = \frac{30-9}{5-2}$$
$$= 7 \ m/s$$

Example 5 discusses **average velocity** because the calculations were done over a **measurable time interval**. Calculus enters the picture when we wish to calculate the velocity at a **specific moment**, or "**instantaneous**" velocity.

To do so we need to "squeeze" the time interval down to a value very close to zero.

TIME INTERVAL	AVERAGE VELOCITY
$2 \leq t \leq 2.1$	4.1
$2 \leq t \leq 2.01$	4.01
$2 \leq t \leq 2.001$	4.001
$2 \leq t \leq 2.0001$	4.0001

As we can see from the table, the instantaneous velocity at 2 seconds APPROACHES 4 m/s.

Similar to the discussion of a tangent at a point, we can now say that the **instantaneous velocity** at time t is **the rate of change of the displacement, as Δt approaches zero**. Thus, velocity is the **derivative of the displacement** with respect to time.

Instantaneous velocity

If $s(t)$ represents the displacement at time t, the velocity is

$$v(t) = s'(t) = \lim_{\Delta t \to 0} \frac{s(t + \Delta t) - s(t)}{\Delta t}$$

Example 6 INSTANTANEOUS VELOCITY

An object falls from a tall building. Its height, in m, after t seconds, is given by the equation:

$$h = 150 - 4.9t^2$$

Determine the instantaneous velocity after (a) 1 second;
(b) 5 seconds.

$$v = \lim_{\Delta t \to 0} \frac{[150 - 4.9(t + \Delta t)^2] - (150 - 4.9t^2)}{\Delta t}$$

$$= \lim_{\Delta t \to 0} \frac{-9.8t(\Delta t) - 4.9(\Delta t)^2}{\Delta t} \qquad \text{\textit{After expanding}}$$
$$\text{\textit{and simplifying}}$$

$$= \lim_{\Delta t \to 0} (-9.8t - 4.9\Delta t)$$

$$= -9.8t$$

(a) At $t = 1$, $v = -9.8$

The velocity is -9.8 m/s.

(b) At $t = 5$, $v = -49$

The velocity is -49 m/s.

NOTE: Negative velocity means that an object is travelling downwards (in this example) or backwards.

MAKING THE CONNECTION BETWEEN SLOPE AND VELOCITY

Slope is defined as the rate of change of y with respect to x. **Velocity** is defined as the **rate of change of the distance with respect to time**. As a result, we can investigate the following diagrams and relate our findings to both slope and velocity.

Each diagram represents a graph of the displacement (vertical axis) of an object, with respect to time (horizontal axis). **Upward** slope means that the object is travelling forward (positive slope, positive velocity). **Downward** slope means it is travelling backwards (negative slope, negative velocity). A **straight line** means that the velocity is constant. A **curved graph** means that the velocity is changing.

24

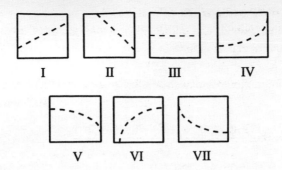

Diagram I -	forward (positive) motion with positive, constant velocity (constant, positive slope) [e.g. a car being driven at a constant speed]
Diagram II -	backward (negative) motion with negative, constant velocity (constant, negative slope) [e.g. a car being driven in reverse at a constant speed]
Diagram III -	stationary object with zero velocity (zero slope) [c.g. a parked car]
Diagram IV -	forward motion with positive, increasing velocity (positive, increasing slope)[e.g. a car being driven forward, accelerating]
Diagram V -	backward motion with negative, decreasing velocity (negative, decreasing slope)[e.g. a car being driven in reverse, accelerating]
Diagram VI -	forward motion with positive, decreasing velocity (positive, decreasing slope) [e.g. a car being driven forward, braking]
Diagram VII -	backward motion with negative, increasing velocity (negative, increasing slope) [e.g. a car being driven in reverse, braking]

Do not be confused with velocity vs. speed. Speed is always a positive quantity. Velocity is directional. Negative velocity means backward motion. Increasing velocity means the change is in the positive direction, whereas decreasing velocity means the change is in the negative direction.

PRACTICE EXERCISE 2

1. Use first principles to find the derivative of each of the following functions.

   (a) $y = 3x^2 + 2$ (b) $y = \sqrt{x-5}$ (c) $y = \dfrac{1}{x^2 - 1}$

2. Determine the slope of the tangent to the curve with equation $y = 6x^2 - 9$ at the point (2,15).
   Graph the curve and the tangent.

3. Determine the slope of the tangent to the curve with equation $y = \dfrac{2}{x^2}$ at the point (-1,2).
   Graph the curve and the tangent.

4. An object has displacement, in metres, given by the equation $s = 2t^3 - 4t$, where t is time in seconds.
   (a) Determine the average velocity during the first 3 s.
   (b) Determine the instantaneous velocity after 4 s.

CHAPTER THREE

Techniques of differentiation

In chapter 2, derivatives by first principles were discussed. This chapter will give hints on the various techniques of differentiation. It will not prove the rules. That will be left to your textbook. Applications will be covered in chapters 4 and 5.

DERIVATIVE OF A POWER

$$\text{If } y = ax^1 \text{ then } \frac{dy}{dx} = nax^{n-1}$$

The derivative of a constant is zero.

Example 1

Find the derivative of $y = 3x^5$

$$\frac{dy}{dx} = 5 \cdot 3x^{5-1}$$

Simple use of the power rule.

$$= 15x^4$$

Example 2

Find the derivative of $f(x) = \dfrac{1}{\sqrt{x}}$

First, rewrite $f(x)$ in exponential form.

$$f(x) = x^{-\frac{1}{2}}$$

$$f^1(x) = -\frac{1}{2}x^{-\frac{1}{2}-1}$$

$$= -\frac{1}{2\sqrt{x^3}}$$

Example 3

Find the derivative of $y=7$

Because y is a constant, $\dfrac{dy}{dx} = 0$

DERIVATIVES OF SUMS AND DIFFERENCES

If $y = f(x) \pm g(x)$

then $\dfrac{dy}{dx} = f'(x) \pm g'(x)$

Example 4

Find the derivative of $y=3x^3+7x^2-2x+6$

$$\frac{dy}{dx} = 3 \cdot 3x^{3-1} + 2 \cdot 7x^{2-1} - 2x^{1-1} + 0$$
$$= 9x^2 + 14x - 2$$

Example 5

Find the derivative of $y=6x^{-5}-7x^{-3}+x^{-1}$

$$\frac{dy}{dx} = -5 \cdot 6x^{-5-1} - (-3)7x^{-3-1} + (-1)x^{-1-1}$$
$$= -30x^{-6} + 21x^{-4} - x^{-2}$$

Take the derivative of each individual term.

Example 6

Determine an equation of the tangent to $y = 2x^3-5x$ at the point $(x,y) = (2,6)$.

To find the slope of the tangent, we find $\dfrac{dy}{dx}$ *at* $(x,y) = (2,6)$, *i.e. at $x = 2$.*

$$\frac{dy}{dx} = 6x^2 - 5$$
$$\left.\frac{dy}{dx}\right|_{x=2} = 6(2)^2 - 5$$
$$= 19$$

Let $\dfrac{y-6}{x-2} = 19$

$y-19x-32$ *is an equation of the tangent.*

PRODUCT RULE

The product rule can be stated in one of two ways:

If $y = u \cdot v$

then $\dfrac{dy}{dx} = \dfrac{du}{dx} \cdot v + \dfrac{dv}{dx} \cdot u$

If $y = f(x) \cdot g(x)$

then $\dfrac{dy}{dx} = f'(x) \cdot g(x) + g'(x) \cdot f(x)$

Example 7

Find the derivative of $y = \left(3x^2 - 5x\right)\left(x^4 + 2\right)$

$Let \quad u = 3x^2 - 5x \quad and \quad v = x^4 + 2$

$\dfrac{dy}{dx} = \dfrac{du}{dx} \cdot v + \dfrac{dv}{dx} \cdot u$

$= (6x - 5)\left(x^4 + 2\right) + 4x^3\left(3x^2 - 5x\right)$

$= 18x^5 - 25x^4 + 12x - 10$

An easy way of remembering the product rule is to take **the derivative of the "first" times the "second" plus the derivative of the "second" times the "first".**

Example 8

Find the slope of the tangent to $y = \sqrt{x}\left(2x^2 - 1\right)$ at the point $(x,y) = (1,1)$.

$$\text{Let}\quad u = \sqrt{x} = x^{\frac{1}{2}} \quad \text{and}\quad v = 2x^2 - 1$$

$$\frac{dy}{dx} = \frac{du}{dx}\cdot v + \frac{dv}{dx}\cdot u$$

$$= \frac{1}{2}x^{-\frac{1}{2}}\left(2x^2 - 1\right) + 4x\cdot x^{\frac{1}{2}}$$

$$= 5x^{\frac{3}{2}} - \frac{1}{2}x^{-\frac{1}{2}}$$

$$\text{Slope} = 5(1)^{\frac{3}{2}} - \frac{1}{2}(1)^{-\frac{1}{2}}$$

$$= \frac{9}{2}$$

QUOTIENT RULE

The Quotient Rule
can be stated in one of two ways:

$$\text{If}\quad y = \frac{u}{v}$$

$$\text{then}\quad \frac{dy}{dx} = \frac{\dfrac{du}{dx}\cdot v - \dfrac{dv}{dx}\cdot u}{v^2}$$

$$\text{If}\quad y = \frac{f(x)}{g(x)}$$

$$\text{then}\quad \frac{dy}{dx} = \frac{f'(x)\cdot g(x) - g'(x)\cdot f(x)}{[g(x)]^2}$$

Example 9

Find the derivative of $y = \dfrac{2x+1}{3x-2}$

$$Let \quad u = 2x+1 \quad v = 3x-2$$

$$\frac{dy}{dx} = \frac{\dfrac{du}{dx} \cdot v - \dfrac{dv}{dx} \cdot u}{v^2}$$

$$= \frac{2(3x-2)-3(2x+1)}{(3x-2)^2}$$

$$= \frac{-7}{(3x-2)^2}$$

Example 10

Find the derivative of $y = \dfrac{2x-3}{x^2-3x+5}$

$$Let \quad u = 2x-3 \quad v = x^2-3x+5$$

$$\frac{dy}{dx} = \frac{\dfrac{du}{dx} \cdot v - \dfrac{dv}{dx} \cdot u}{v^2}$$

$$= \frac{2(x^2-3x+5)-(2x-3)(2x-3)}{(x^2-3x+5)^2}$$

$$= \frac{2x^2-6x+10-4x^2+12x-9}{(x^2-3x+5)^2}$$

$$= \frac{-2x^2+6x+1}{(x^2-3x+5)^2}$$

An easy way of remembering the quotient rule is to take **the derivative of the "top" times the "bottom" minus the derivative of the "bottom" times the "top", over the square of the "bottom"**. Notice that the order of the derivatives is the same as with the Product Rule. The addition has been changed to subtraction.

Example 11

Find the equation of the tangent to

$$y = \frac{\sqrt{x}}{\sqrt{x}-1} \quad \text{at the point } (x,y) = (4,2).$$

Let $\quad u = \sqrt{x} = x^{\frac{1}{2}} \quad v = \sqrt{x}-1 = x^{\frac{1}{2}}-1$

$$\frac{dy}{dx} = \frac{\frac{du}{dx}\cdot v - \frac{dv}{dx}\cdot u}{v^2}$$

$$= \frac{\frac{1}{2}x^{-\frac{1}{2}}\left(x^{\frac{1}{2}}-1\right) - \frac{1}{2}x^{-\frac{1}{2}}\left(x^{\frac{1}{2}}\right)}{\left(x^{\frac{1}{2}}-1\right)^2}$$

$$= \frac{\frac{1}{2}x^0 - \frac{1}{2}x^{-\frac{1}{2}} - \frac{1}{2}x^0}{\left(x^{\frac{1}{2}}-1\right)^2}$$

$$= \frac{-1}{2\sqrt{x}\left(\sqrt{x}-1\right)^2}$$

At $(x,y) = (4,2)$

$$Slope = \frac{-1}{2\sqrt{4}\left(\sqrt{4}-1\right)^2}$$

$$= -\frac{1}{4}$$

$$\frac{y-2}{x-4} = -\frac{1}{4}$$

$$x + 4y - 12 = 0 \quad \textit{is the tangent equation.}$$

The Chain Rule can be stated in either of two ways:

If $y = f(g(x))$

Then $\dfrac{dy}{dx} = f'(g(x)) \times g'(x)$

If $y = f(g(x))$

Let $u = g(x)$ so that $y = f(u)$

Then $\dfrac{dy}{dx} = \dfrac{dy}{du} \times \dfrac{du}{dx}$

Example 12

Find the derivative of $y = (x^2 - 7)^5$.

Using the chain rule, let $u = x^2 - 7$

$$So \quad y = u^5$$

$$\frac{dy}{dx} = \frac{dy}{du} \times \frac{du}{dx}$$

$$Now, \quad \frac{dy}{du} = 5u^4 = 5(x^2 - 7)^4$$

$$And \quad \frac{du}{dx} = 2x$$

$$\therefore \frac{dy}{dx} = 5(x^2 - 7)^4 2x$$

$$= 10x(x^2 - 7)^4$$

Example 13

Find the derivative of $y = \dfrac{2}{\sqrt{1-x^3}}$.

$$y = 2(1-x^3)^{-\frac{1}{2}}$$

$$Let \quad u = 1-x^3$$

$$\frac{dy}{dx} = \frac{dy}{du} \times \frac{du}{dx}$$

$$= 2\left(-\frac{1}{2}\right)(1-x^3)^{-\frac{3}{2}}(-3x^2)$$

$$= 3x^2(1-x^3)^{-\frac{3}{2}}$$

A common mistake is to use the quotient rule here. This is unnecessary because of the constant in the numerator. The chain rule is appropriate because the function can be expressed in exponential form.

A good way of remembering how to apply the chain rule is to take the **derivative of the "outer" function times the derivative of the "inner" function.**

Example 14

Find the slope of the tangent to $y = \left(\dfrac{4x-5}{3x+2}\right)^4$ at $x = 1$.

This example combines the chain rule with the quotient rule. Be careful to apply the chain rule first.

Let $u = \dfrac{4x-5}{3x+2}$

$\dfrac{dy}{dx} = \dfrac{dy}{du} \times \dfrac{du}{dx}$

$\quad = 4\left(\dfrac{4x-5}{3x+2}\right)^3 \dfrac{4(3x+2)-3(4x-5)}{(3x+2)^2}$ *Apply the chain rule first, followed by the quotient rule.*

$\quad = \dfrac{4(4x-5)^3(23)}{(3x+2)^5}$ *Simplify both the numerator and denominator by collecting like*

$\quad = \dfrac{92(4x-5)^3}{(3x+2)^5}$ *terms.*

At $x = 1$,

$\dfrac{dy}{dx} = \dfrac{92[4(1)-5]^3}{[3(1)+2]^5}$

$\quad = \dfrac{-92}{3125}$

$\therefore Slope = \dfrac{-92}{3125}$

IMPLICIT DIFFERENTIATION

Generally, functions are written **explicitly,** in the form $y=f(x)$. The x terms go on the one side of the equation and the y term goes alone on the other side. But, many relations are written **implicitly,** with **y embedded in the expression**. Here are three such examples:

$x^2+y^2=9$ $3x^2-7xy+8y^2-16=0$ $y^2=3x$

This section will illustrate a method of finding the derivative, by use of the chain rule, when written implicitly.

Recall that, if $y = u^2$

then $\dfrac{dy}{dx} = 2u\dfrac{du}{dx}$

Example 15

Find the derivative of $y^2 = x$.

$$\dfrac{dy^2}{dx} = \dfrac{dx}{dx}$$ *Apply the chain rule to the y^2 term.*

$$2y\dfrac{dy}{dx} = 1$$

$$\dfrac{dy}{dx} = \dfrac{1}{2y}$$

Example 16

Find the derivative of $x^2 + xy - y^2 = 4$.

This relation combines the chain rule and the product rule.

$$\dfrac{dx^2}{dx} + \dfrac{d(xy)}{dx} - \dfrac{dy^2}{dx} = \dfrac{d4}{dx}$$ *Take the derivative of each term, applying the chain rule to the y^2 term.*

$$2x + (\dfrac{dx}{dx}y + \dfrac{dy}{dx}x) - 2y\dfrac{dy}{dx} = 0$$

$$2x + 1y + x\dfrac{dy}{dx} - 2y\dfrac{dy}{dx} = 0$$ *Use the product rule on the xy term.*

$$(x - 2y)\dfrac{dy}{dx} = -y - 2x$$ *Collect like terms and simplify.*

$$\dfrac{dy}{dx} = \dfrac{-y - 2x}{x - 2y}$$

Example 17
Find an equation of the tangent to $x^2 + y^2 = 289$ at the point $(x,y) = (8,15)$.

Use implicit differentiation first

$$2x + 2y\frac{dy}{dx} = 0$$

$$2y\frac{dy}{dx} = -2x$$

$$\frac{dy}{dx} = \frac{-2x}{2y}$$

$$\frac{dy}{dx} = \frac{-x}{y}$$

$$At\ (x,y) = (8,15)$$

$$\frac{dy}{dx} = \frac{-8}{15}$$

*Use point – slope form to find
the tangent equation:*

$$\frac{y-15}{x-8} = \frac{-8}{15}$$
$$8x + 15y - 161 = 0$$

DERIVATIVES OF HIGHER ORDER

As we know, if the function $y = f(x)$ is differentiable, then its derivative is symbolized by one of:

$$y' \qquad \frac{dy}{dx} \qquad f'(x) \qquad D_x y$$

If this, in turn, is a differentiable function, then we may find what is known as the **second derivative** of y with respect to x, or:

$$y'' \qquad \frac{d^2y}{dx^2} \qquad f''(x) \qquad D_x^2 y$$

37

This can be interpreted as the change in the slope (to be dealt with in chapter 5) or the change in velocity, which is called acceleration (to be dealt with in chapter 4).

Example 18

Find the second derivative of $y = 5x^4 - 7x^3 + 3x^2 + 6x - 9$

$$\frac{dy}{dx} = 20x^3 - 21x^2 + 6x + 6$$

$$\frac{d^2y}{dx^2} = 60x^2 - 42x + 6$$

Example 19

Find $f''(x)$ for the function $f(x) = (3x^2 - 1)^5$.

$$f'(x) = 5(3x^2 - 1)^4(6x)$$
$$= 30x(3x^2 - 1)^4$$

Use the Chain rule

$$f''(x) = 30(3x^2 - 1)^4 + 30x \cdot 4(3x^2 - 1)^3(6x)$$
$$= 30(3x^2 - 1)^4 + 720x^2(3x^2 - 1)^3$$
$$= \left[30(3x^2 - 1) + 720x^2\right](3x^2 - 1)^3$$
$$= (810x^2 - 30)(3x^2 - 1)^3$$
$$= 30(27x^2 - 1)(3x^2 - 1)^3$$

Use the Product rule and then the Chain rule for $(3x^2-1)^3$

Common factor of $(3x^2-1)^3$

Example 20

Find $\dfrac{d^2y}{dx^2}$ for the equation $x^2 - y^2 = 9$.

Begin with implicit differentiation

$$2x - 2y\frac{dy}{dx} = 0$$

$$-2y\frac{dy}{dx} = -2x$$

$$\frac{dy}{dx} = \frac{x}{y}$$

Use the quotient rule, with implicit differentiation for the y term

$$\frac{d^2y}{dx^2} = \frac{1 \cdot y - \frac{dy}{dx} \cdot x}{y^2}$$

$$= \frac{y - \left(\frac{x}{y}\right) \cdot x}{y^2} \qquad \text{Substitute } \tfrac{x}{y} \text{ for } \tfrac{dy}{dx}$$

$$= \frac{y - \frac{x^2}{y}}{y^2} \qquad \text{Common denominator of } y$$

$$= \frac{\frac{y^2 - x^2}{y}}{y^2}$$

$$= \frac{-9}{y^3} \qquad \text{Given } x^2 - y^2 = 9 \quad \therefore \quad y^2 - x^2 = -9$$

Note, problem solving is not limited to first and second derivatives. Third, fourth, etc., derivatives may be found in the same manner.

PRACTICE EXERCISE 3

1. Determine the derivative of each of the following. Simplify fully.

   (a) $y = 5x^3 - 4x^2 + 7x + 3$

   (b) $f(x) = 8x^{-2} + 4x^{-3}$

   (c) $y = 3x\sqrt{x-6}$

   (d) $h(x) = (3x^4 - 5x)^6$

   (e) $y = \dfrac{2x}{x^2 + 1}$

   (f) $y = (3x+2)(x+5)^4$

   (g) $k(x) = x^2(2x^3 + x)^{\frac{1}{3}}$

   (h) $y = \left(\dfrac{x^2 + 3x}{2x + 3}\right)^3$

2. Determine the first, second and third derivatives of $y = \dfrac{6}{x} - \dfrac{3}{x-2}$

3. Find $\frac{dy}{dx}$ for each of the following. Then find an equation of the tangent at the point given.

   (a) $x^2 + y^2 = 34$ at $(-3,5)$

   (b) $2x^3 - 8xy + y^2 = 1$ at $(2,1)$

CHAPTER FOUR

Derivatives and rates of change

A ratio is the comparison of two quantities with the same units of measurement. A rate is the comparison, in ratio form, when the units of measurement are different. Examples of these units of measurement would include km/h, kg/m, \$/hamburger, etc. This chapter will investigate a variety of examples involving rates of change, giving hints and ideas on techniques of solving for rate problems, especially ones in which the instantaneous rate of change needs to be found.

VELOCITY AND ACCELERATION

Remember that velocity is the rate of change of the displacement (distance from a fixed point) with respect to time. To find the instantaneous velocity, we find $\frac{ds}{dt}$.

Example 1 DISPLACEMENT & VELOCITY

The position (in metres) of an object after t seconds is given by the equation

$$s(t) = t^2 - 5t + 8, \ t \geq 0$$

(a) Find the velocity after t seconds.

To find the velocity, we simply find s'(t)
$$v = s'(t) = 2t - 5$$

(b) When is the object at rest?

An object is at rest when its velocity is zero.
$$Let \ v = 0$$
$$2t - 5 = 0$$
$$t = 2.5$$
The object is at rest at 2.5 s.

41

(c) When is the object moving in a positive direction?

For an object to move in a positive direction,
its velocity must be positive.

$$Let \; v > 0$$
$$2t - 5 > 0$$
$$t > 2.5$$

The velocity is positive after 2.5 s.

 Acceleration, by definition, is the **rate of change of the velocity** with respect to time. Thus, to find the instantaneous acceleration, we simply find $\frac{dv}{dt}$ or $\frac{d's}{dt^2}$.

Example 2
DISPLACEMENT & ACCELERATION

The position, in metres, of an object after t seconds is given as
$$s(t) = 5t^3 - 3t^2 + t - 6.$$

(a) Find the acceleration of the object after t seconds.

To find the acceleration, we must first find $v = s'(t)$ and then $a = v'(t)$

$$v = s'(t) = 15t^2 - 6t + 1$$
$$a = v'(t) = 30t - 6$$

(b) When is the acceleration zero?

$$Let \; a = 0$$
$$30t - 6 = 0$$
$$t = 0.2$$

The acceleration is zero at 0.2 s.

(c)   When is the acceleration positive?

$$Let \ a > 0$$
$$30t - 6 > 6$$
$$t > 0.2$$

The acceleration is positive after 0.2 s.

(d)   When is the acceleration negative?

$$Let \ a < 0$$
$$30t - 6 < 6$$
$$t < 0.2$$
The acceleration is negative before 0.2 s.

 Negative acceleration means that the velocity is decreasing. (Velocity can have any value.)

Example 3
GRAVITATIONAL ACCELERATION

A ball is tossed into the air. Its height, in metres above the ground after t seconds, is given by the equation:

$$h \ (t) = -4.9t^2 + 5t + 1.5, \ t \ge 0$$

(a)   What is the initial height of the ball?
To find initial quantities, we always let $t=0$.

$$h \ (0) = -4.9(0)^2 + 5(0) + 1.5$$
$$= 1.5$$

The initial height of the ball is 1.5m

43

(b)   How long does it take for the ball to fall to the ground?
       When the ball falls to the ground, the height is zero.

$$Let \quad -4.9t^2 + 5t + 1.5 = 0$$

$$t = \frac{-5 \pm \sqrt{5^2 - 4(-4.9)(1.5)}}{2(-4.9)} \qquad \left[t = \frac{-b \pm \sqrt{b^2 - 4ac}}{2a} \right]$$

$$\doteq -0.2424 \; or \; 1.2628$$

$$t \geq 0 \quad \therefore t \neq -0.2424$$

$$\therefore t \doteq 1.2628$$

The ball falls to the ground after about 1.26 s.

(c)   What is the initial velocity?

$$v(t) = h'(t) = -9.8t + 5$$
For initial velocity, let $t = 0$
$$v(0) = -9.8(0) + 5$$
$$= 5$$

The initial velocity was 5 m/s.

(d)   What is the acceleration due to gravity?

$$a = v'(t) = -9.8 \; m / s^2$$

 It is important to remember that, for all objects falling freely, acceleration due to gravity is approximately $-9.8 m/s^2$.

APPLICATIONS IN THE SCIENCES

Aside from velocity and acceleration, there are many applications in the sciences. In this section, applications to the physical sciences will be looked at. Applications in biology will be seen in Chapter 8, Exponential and Logarithmic Functions.

Example 4
STOPPING DISTANCE

The average distance, in metres, required to stop a car traveling at x km/h is given as $y = 0.005x^2$.

Determine the rate of change of the stopping distance when
(a) $x = 40$ km/h

$$y = 0.005x^2$$

$$\frac{dy}{dx} = 0.01x$$

$$At\ x = 40,\ \frac{dy}{dx} = 0.01(40)$$

$$= 0.4$$

The rate of change of the stopping distance is 0.4m/ km/h.

(b) $x = 100$ km/h

$$\frac{dy}{dx} = 0.01x$$

$$At\ x = 100,\ \frac{dy}{dx} = 0.01(100)$$

$$= 1$$

The rate of change of the stopping distance is 1m/ km/h.

The rate of change of the stopping distance represents how much the stopping distance is changing, per km/h change in speed, at that moment. Note that both the stopping distance and the rate of change of the stopping distance increase as the speed increases. This is due to the quadratic equation used in this model.

You can also see that the rate of change here is not with respect to time, as it was with velocity and acceleration. Unless direct instructions, or clues are given to differentiate with respect to time, simply use the variables present in the model. Ideas on differentiating with respect to time, when time is not part of the model, are discussed under **Related Rates**, later in this chapter.

Example 5
ILLUMINATION

The intensity of illumination, in candles, on the ground of a light source, D metres above the ground, is given by the equation

$$I = \frac{k}{D^2}$$

The intensity is 200 cd when the light is 3 m above the ground. What is the rate of change of intensity when the light is 1 m from the ground?

$I = \dfrac{k}{D^2}$

Substitute I = 200 and D = 3

$200 = \dfrac{k}{3^2}$

$k = 1800$

$\therefore I = \dfrac{1800}{D^2} = 1800D^{-2}$

$\dfrac{dI}{dD} = -3600D^{-3}$

At $D = 1, \dfrac{dI}{dD} = -3600(1)^{-3}$

$= -3600$

The intensity of illumination is changing at a rate of −3600 cd/m.

Example 6
PENDULUM

The period of a pendulum is the time taken for one complete swing, back and forth. The period, in seconds, is given by the equation

$$T = 2\pi\sqrt{\frac{L}{9.8}}$$

where L is the length of the pendulum, in metres. Determine the rate of change of the period when the length of the pendulum is 2 m.

$$T = 2\pi\sqrt{\frac{L}{9.8}}$$

$$= 2.007 L^{\frac{1}{2}}$$

$$\frac{dT}{dL} = 1.0035 L^{-\frac{1}{2}}$$

$$At\ L = 2, \frac{dT}{dL} = 1.0035(2)^{-\frac{1}{2}}$$

$$= 0.7096$$

The rate of change of the period is about 0.71 s/m.

APPLICATIONS IN THE SOCIAL SCIENCES

There are also numerous applications in the social sciences. Most frequently, although not exclusively, we see rates of change applications in Demography and Economics.

Example 7
POPULATION GROWTH

A model to predict the population of a town, after t years, was developed. The following equation was provided:

$$P(t)\quad = 10\ 000 + 100t^2 - 5t^3 \quad 0 \le t \le 20$$

Determine the population and the rate of change of the population after 15 years.

$$P(15)\quad = 10\ 000 + 100(15)^2 - 5(15)^3$$
$$= 15625$$

The predicted population will be 15 625 people.

$$P'(t)\quad = 200t - 15t^2$$
$$P'(15)\quad = 200(15) - 15(15)^2$$
$$= -375$$

The population is predicted to be declining
at a rate of 375 people/year after 15 years.

In the next example, the term **Marginal Cost** is used. To economists, **marginal** means the rate of change. Marginal Cost would then be the change in cost for one more unit of the product produced. This can be represented in Calculus by rates of change methods.

Example 8
MARGINAL COST

A manufacturer estimates the cost, in dollars, of making x light bulbs to be:

$$C(x) = 10\,000 + 10\sqrt{x}$$

What is the marginal cost of making 100 000 light bulbs?

$$C(x) = 10\,000 + 10x^{\frac{1}{2}}$$

$$C'(x) = 5x^{-\frac{1}{2}}$$

The marginal cost function can be found by differentiating C with respect to x.

$$C'(100\,000) = 5(100\,000)^{-\frac{1}{2}}$$
$$= 0.01581$$

The marginal cost is about $0.016 per light bulb.

RELATED RATES

Related rates problems generally provide for an equation in one set of quantities, such as distance or volume, but request rates of change with respect to another quantity, usually time. The rate of change of one quantity is usually specified and we are asked to find the rate of change of another, but related, quantity. Thus, the term, "Related Rates".

48

Techniques for solving related rates problems:

- Picture, in your mind, the event occurring, instead of looking at it as a math problem. This will help you draw a diagram. The event will usually be described in the first one or two sentences.
- Add the dimensions to your diagram and label the unknown quantities.
- If a quantity is not constant, (e.g., "The length is 2 m *when* the height is 1.5 m"), do not use it to label the diagram. Write it separately at the top of your solution in mathematical terms.
- Define a set of variables (e.g., "Let *x* represent the distance from the boat to the shore." If the rate of change is with respect to time, don't forget to define your variable for time, as well.
- Write an equation relating the variables. (Usually only 1 equation is required.) Use the diagram to assist you. Generally, the information regarding rates of change are not used here.
- Differentiate with respect to the appropriate quantity. This quantity is usually *time*. In most problems, the chain rule will be used.
- Resist the urge to substitute non-constant values until now. Decide whether you need to substitute into the original equation first, based on the given clues (e.g., "The length is 2 m *when* the height is 1.5 m").
- Solve for the requested unknown quantity and evaluate at the given values.
- Provide a concluding statement, in sentence form, appropriately answering the original question. (Unfortunately, this is where many students' beautiful solutions fall apart.)

The following example is a typical **area** problem. The situation is of a circular area that is growing over a period of time. The equation involves area and distance, but the problem involves a rate of change with respect to time.

Example 9
CHANGE IN AREA

When a pebble was dropped into a pool of water, it produced a circular wave that traveled outward at a constant speed of 10 cm/s. At what rate is the area inside the wave increasing when its radius is 15 cm?

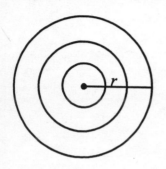

The ripple is circular. Therefore the area is calculated by the formula for the area of a circle.

Note that the radius is not constant ("when its radius is 15 cm")

Let **r** represent the radius in cm.
Let *A* represent the area in cm².
Let *t* represent the time in s.

Given $\dfrac{dr}{dt} = 10$

When $r = 15$, $\dfrac{dA}{dt} = ?$

$A = \pi r^2$ *Area of a circle*

$\dfrac{dA}{dt} = 2\pi r \dfrac{dr}{dt}$ *Differentiate with*
respect to time to
Substitute $r = 15$ and $\dfrac{dr}{dt} = 10$ *get rate of change*
of the radius
$\dfrac{dA}{dt} = 2\pi(15)(10)$ *and of the area.*

$= 942.48$

The area is increasing at a rate of about 942.48 cm²/s

Example 10
RIGHT TRIANGLE

A ladder 3 m long is sliding down a wall at 15 cm/s. How fast is the bottom of the ladder sliding outward when the top is 1 m from the ground?

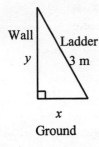

The ladder, ground and wall form a right-angled triangle.
Therefore, we use the Pythagorean Theorem.

Only the length of the ladder is constant. The other measurements occur only at a specific moment.

Let x represent the distance, in m, from the wall to the base of the ladder.
Let y represent the distance, in m, from the ground to the top of the ladder.
Let t represent the time, in s.

$$Given \ \frac{dy}{dt} = -15 \, cm/s = -0.15 \, m/s$$

$(\frac{dy}{dt}$ *is negative because y is getting smaller*$)$

When y $= 1, \ \frac{dx}{dt} = ?$ *Pythagorean Theorem*

$$x^2 + y^2 = 3^2$$
$$x^2 + y^2 = 9$$
$$2x\frac{dx}{dt} + 2y\frac{dy}{dt} = 0$$ *Differentiate (implicitly) with respect to t.*
$$\frac{dx}{dt} = -\frac{y}{x}\frac{dy}{dt}$$

$$\text{At } y = 1, \quad x^2 + 1^2 = 9$$
$$x^2 = 8$$
$$x = \sqrt{8}$$

Substitute for y so as to find the horizontal distance, x, at this moment.

Substitute $x = \sqrt{8}$, $y = 1$ and $\dfrac{dy}{dt} = -0.15$

$$\frac{dx}{dt} = -\frac{1}{\sqrt{8}}(-0.15)$$
$$= 0.053$$

The bottom of the ladder is sliding outward at a rate of about 0.053 m/s or 5.3 cm/s.

Example 11
SIMILAR TRIANGLES

A lamppost is 5 m high. A man of height 1.8 m passes under the lamppost, walking at 3 m/s.
(a)   At what rate is the length of his shadow increasing?
(b)   How fast is the end of his shadow moving?

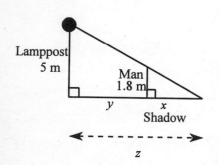

Although this is a right-angled triangle, the Pythagorean Theorem is not used here because there is no reference to the hypotenuse.

The diagram contains similar triangles, so proportions can be used.

The heights of both the lamppost and the man are constants.

*Let **x** represent the length, in m, of the shadow.*
*Let **y** represent the distance, in m, from the lamppost to the man.*
*Let **z** represent the distance, in m, from the lamppost to the end of the shadow.*
*Let **t** represent the time, in s.*

52

Given $\dfrac{dy}{dt} = 3\,m/s$

(a) $\quad \dfrac{dx}{dt} = ?$

Using proportions,

$$\dfrac{x}{x+y} = \dfrac{1.8}{5}$$

$$5x = 1.8x + 1.8y$$

$$3.2x = 1.8y$$

$$x = 0.5625y$$

$$\dfrac{dx}{dt} = 0.5625\dfrac{dy}{dt}$$

Substitute $\dfrac{dy}{dt} = 3$

$$\dfrac{dx}{dt} = 0.5625(3)$$

$$= 1.6875$$

The length of his shadow
is increasing at a rate of
about 1.7 m/s

(b) $\quad \dfrac{dz}{dt} = ?$

From the diagram, we can see that

$$z = x + y$$

$$\dfrac{dz}{dt} = \dfrac{dx}{dt} + \dfrac{dy}{dt}$$

$$\dfrac{dz}{dt} = 1.6875 + 3 \qquad [\dfrac{dx}{dt} = 1.6875 \; from \; (a)]$$

$$\dfrac{dz}{dt} = 4.6875$$

The end of his shadow is moving at a
speed of about 4.7 m/s

Example 12
VOLUME OF A SOLID

A spherical balloon is being pumped with air at a rate of 10 cm³/s.
How fast is the radius increasing when the radius is 5 cm?

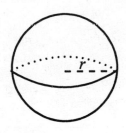

**We will use the formula for the volume of
a sphere.**

**There are no constant dimensions here.
The radius is substituted after finding the
derivative.**

53

*Let **r** represent the radius, in cm*
*Let **V** represent the volume, in cm³*
*Let **t** represent the time, in s.*

$$\text{Given} \quad \frac{dV}{dt} = 10$$

$$At \quad r = 5, \quad \frac{dr}{dt} = ?$$

$$V = \frac{4}{3}\pi r^3 \qquad\qquad \textit{Volume of a sphere}$$

$$\frac{dV}{dt} = 4\pi r^2 \frac{dr}{dt}$$

$$\frac{dr}{dt} = \frac{1}{4\pi r^2}\frac{dV}{dt} \qquad \textit{Differentiate with respect to time}$$

Substitute $r = 5$ and $\dfrac{dV}{dt} = 10$

$$\frac{dr}{dt} = \frac{1}{4\pi(5)^2}(10)$$

$$= 0.03183$$

The radius is increasing at a rate of about 0.03 cm/s

A final hint when substituting for the given rate of change...

Be conscious of the direction of change. If a quantity is increasing in value, use a positive value for the rate of change (Examples 9, 11 and 12). If the quantity is decreasing in value, use a negative value for the rate of change (Example 10).

PRACTICE EXERCISE 4

1. A particle moves so that its position is given by the function

$$s = 4t^3 - 8t^2 + t - 50$$

where t is in minutes and s is in metres.
(a) Find the velocity and acceleration after t minutes.
(b) Find the velocity and acceleration after 5 minutes.
(c) When will the velocity be zero?

2. Given the position function $s(t) = \sqrt{12t - 8}$ where s is in metres and t is in seconds. Determine the acceleration after 6 seconds.

3.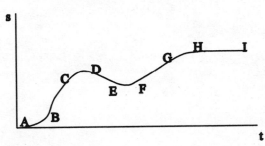

Given the graph above of the displacement of an object, describe the velocity and acceleration in each interval.

4. The population of a species of fish is given by the equation

$P(t) = 1.2 \times 10^8 (1 + 0.02t)^{-2}$ where t is time in years. What is the rate of change of the fish population after 5 years?

5. A company estimates that its profit, in dollars, for x items to be represented by the equation

$$P(x) = 0.005x^2 - 10x - 50\,000$$

Determine the profit and the marginal profit when 100 000 items are sold.

6. A spherical weather balloon has a radius of 1 m when it is 1500 m high. You observe that the radius increases at a rate of 2 cm/min. as it continues to rise. At what rate is the surface area increasing when the radius is 4 m?

7. A passenger car is approaching a railway crossing from 105 m to the east, at 54 km/h. Meanwhile a train approaches from 140 m to the north, at 72 km/h. How fast is the distance between them changing after 5 seconds?

8. Water is pouring into a conical tank at a rate of 20 m³/min. At the top of the tank, the height is 15 m and the radius is 3 m. At what rate is the water level rising when the depth is 2 m?

CHAPTER FIVE

Graphing techniques

LOCAL MAXIMUM AND MINIMUM VALUES

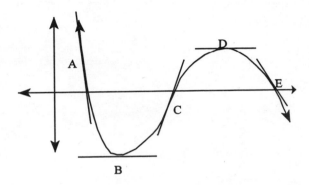

Using the graph above, the slopes at the given points are discussed below.

POINT	DESCRIPTION	DERIVATIVE
A	Decreasing y-values ∴ negative slope	$\dfrac{dy}{dx} < 0$
B	No change in y values ∴ zero slope	$\dfrac{dy}{dx} = 0$
C	Increasing y-values ∴ positive slope	$\dfrac{dy}{dx} > 0$
D	No change in y values ∴ zero slope	$\dfrac{dy}{dx} = 0$
E	Decreasing y-values ∴ negative slope	$\dfrac{dy}{dx} < 0$

The points B and D are both known as **Stationary Points** because the slope of the tangent is zero.

The stationary point at B is a **Local Minimum**, because the neighbouring slope changes from negative to zero to positive.

The stationary point at D is a **Local Maximum**, because the neighbouring slope changes from negative to zero to positive.

Example 1
Given the equation $y=3x^2-1$.
(a)   Discuss the slope at and near any stationary points. Determine the intervals in which the graph is increasing and/or decreasing. Determine any local maximum or minimum points.
(b)   Graph the curve.

(a)   *To be able to discuss the slope, we need to first find the derivative.*

$$\frac{dy}{dx} = 6x$$

Let $\quad \frac{dy}{dx} = 0 \qquad$ (for stationary points)

$$6x = 0$$
$$x = 0$$

At $x = 0, y = 3(0)^2 - 1 = -1$

$\therefore (x,y) = (0,-1)$ *is a stationary point.*

If $x < 0, \dfrac{dy}{dx} < 0$ *i.e., negative slope*

$\therefore$ *For x < 0, the function is decreasing.*

If $x > 0, \dfrac{dy}{dx} > 0$ *i.e., positive slope*

$\therefore$ *for x > 0, the function is increasing.*

$\therefore (x,y) = (0,-1)$ *is a local minimum.*

(b)

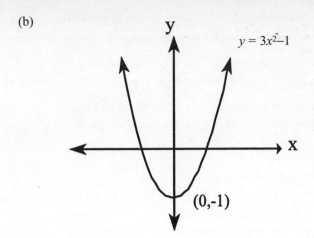

$y = 3x^2 - 1$

y

x

(0,-1)

Example 2

Given the equation $y = x^3 + 3x^2 + 1$.

Determine the local maximum and minimum points. Then graph the curve.

(a) $\dfrac{dy}{dx} = 3x^2 + 6x$

$= 3x(x + 2)$

$Let\ \dfrac{dy}{dx} = 0$

$3x(x + 2) = 0$

$x = 0\quad or\quad x = -2$

$y = 1\qquad y = 5$

The points (0.1) and (-2,5) are stationary points.

Set up a table of values to investigate increasing and decreasing intervals.

INTERVAL	x	x +2	$\dfrac{dy}{dx}$	y
x < -2	-	-	+	increasing
-2 < x < 0	-	+	-	decreasing
x > 0	+	+	+	increasing

59

Based on the increasing and decreasing intervals, we can now say that (0,1) is a local minimum and (−2,5) is a local maximum.

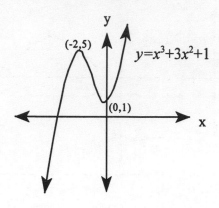

$$y=x^3+3x^2+1$$

CONCAVITY

You'll remember convex and concave lenses from science class. Convex lenses curve outward, whereas concave lenses curve inward.

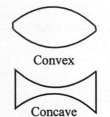

Convex

Concave

Curves, in Calculus, are known only by their **concavity**.

Curves that open **upward** are known as **concave up** and those opening **downward** are known as **concave down**.

Consider the graph below. Consider, also, that the **change in slope** is represented by the **second derivative**.

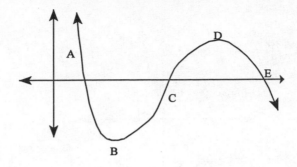

INTERVAL	CONCAVITY	DESCRIPTION	SECOND DERIVATIVE
A to C	Up	Slope changing from negative to zero to positive	$\dfrac{d^2y}{dx^2} > 0$
C to E	Down	Slope changing from positive to zero to negative	$\dfrac{d^2y}{dx^2} < 0$

The above illustration shows that a stationary point is a local minimum if $\dfrac{d^2y}{dx^2} > 0$ and is a local maximum if $\dfrac{d^2y}{dx^2} < 0$.

Second derivative test for maxima and minima

If $\dfrac{dy}{dx} = 0$, then if $\dfrac{d^2y}{dx^2} > 0$, the point is a local minimum

but if $\dfrac{d^2y}{dx^2} < 0$, the point is a local maximum.

Example 3

Given the equation from Example 2: $y = x^3 + 3x^2 + 1$.

Use the second derivative test to determine the local maximum and minimum points.

$$\frac{dy}{dx} = 3x^2 + 6x$$

$$\frac{d^2y}{dx^2} = 6x + 6$$

$Let \ \frac{dy}{dx} = 0$ **For stationary points**

$$3x^2 + 6x = 0$$

$$3x(x + 2) = 0$$

$$x = 0 \ or \ x = -2$$

$$y = 1 \quad y = 5$$

$$At \ x = 0, \quad \frac{d^2y}{dx^2} = 6(0) + 6$$

$$= 6$$

$\frac{d^2y}{dx^2} > 0 \quad \therefore Concave \ up$ **Second derivative test**

$\therefore (0,1) \ is \ a \ local \ minimum \ point.$

$$At \ x = -2, \quad \frac{d^2y}{dx^2} = 6(-2) + 6$$

$$= -6$$

$$\frac{d^2y}{dx^2} < 0 \quad \therefore Concave \ down$$

$\therefore (-2,5) \ is \ a \ local \ maximum \ point.$ **See example 2**

Example 4

Determine all local maximum and minimum points and sketch the graph of $y = x^4 - 4x^3 - 2x^2 + 12x + 1$.

$$\frac{dy}{dx} = 4x^3 - 12x^2 - 4x + 12$$

$$\frac{d^2y}{dx^2} = 12x^2 - 24x - 4$$

For stationary points, let $\frac{dy}{dx} = 0$

$$4x^3 - 12x^2 - 4x + 12 = 0$$
$$4(x-1)(x+1)(x-3) = 0$$

$x = 1$	*or*	$x = -1$	*or*	$x = 3$
$y = 8$		$y = -8$		$y = -8$
$\frac{d^2y}{dx^2} = -16$		$\frac{d^2y}{dx^2} = 32$		$\frac{d^2y}{dx^2} = 32$
< 0		> 0		> 0
Concave down		Concave up		Concave up

$\therefore$ *Local maximum at* $(1,8)$

and local minima at $(-1,-8)$ *and* $(3,-8)$

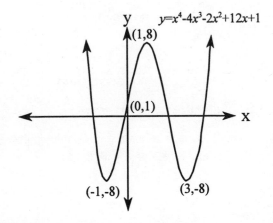

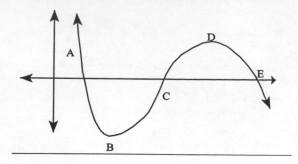

Consider, again the diagram above. At point C, the concavity is changing from upward to downward. *Point C* is called a *POINT OF INFLECTION*. Note, also, that the *second derivative* changes from *positive to negative*. You can see that, to make such a change, the second derivative must be zero at some point. That point is the inflection point.

Second derivative test for points of inflection

If $\dfrac{d^2y}{dx^2} = 0$ and is changing sign

(i.e., concavity is changing), that point is a point of inflection.

Example 5

Use the first and second derivatives to find all maximum, minimum and inflection points, and graph the curve with equation

$$y = x^3 + 3x^2 - 9x - 10.$$

$$\frac{dy}{dx} = 3x^2 + 6x - 9$$

$$\frac{d^2y}{dx^2} = 6x + 6$$

For maxima and minima, let $\dfrac{dy}{dx} = 0$

$3x^2 + 6x - 9 = 0$

$3(x + 3)(x - 1) = 0$

$x = -3 \quad$ or $\quad x = 1$ **Substitute to find y**

$y = 17 \qquad\qquad y = -15$ **Substitute to find**

$\dfrac{d^2y}{dx^2} = -12 \qquad \dfrac{d^2y}{dx^2} = 12$ $\dfrac{d^2y}{dx^2}$

$\quad < 0 \qquad\qquad\quad > 0$

Concave   *Concave*

Downward   *Upward*

A local maximum at (-3,17)

and a local minimum at (1,-15).

For points of inflection, let $\dfrac{d^2y}{dx^2} = 0$

$6x + 6 = 0$

$\quad x = -1$

$\quad y = 1$

$\dfrac{d^2y}{dx^2}$ *is changing from negative to positive.*

$\therefore$ *Point of inflection at* $(-1,1)$.

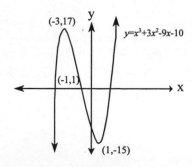

65

It is important to remember that the **concavity must be changing** before a point can be called a point of inflection. Consider a straight line, with equation $y=2x-3$. Any straight line has no concavity because it is "straight". However:

$$\frac{dy}{dx} = 2$$

$$\frac{d^2y}{dx^2} = 0$$

Note that the second derivative is zero, but the first derivative is 2, a constant. Thus, there is no point of inflection.

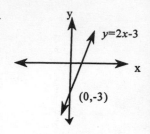

ASYMPTOTES

Many functions are discontinuous, i.e., the graph contains gaps or points at which the function does not exist. Frequently, a graph will approach a straight line, without crossing it, at infinity. This line is known as an **asymptote**.

The graph of $y = \dfrac{1}{x}$ is well known. We will use it to develop methods of finding equations of asymptotes.

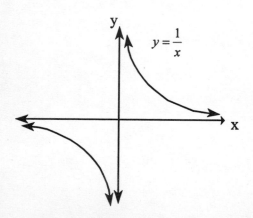

Notice that there is a vertical asymptote along the y-axis, which has equation $x=0$, and a horizontal asymptote along the x-axis, which has equation $y=0$. Neither of these asymptotes is crossed at infinity, but the curve approaches them.

VERTICAL ASYMPTOTES

In rational functions, the equation of the vertical asymptote is found by **setting the denominator equal to zero**. As a result, for $y=\frac{1}{x}$ the equation of the vertical asymptote is $x=0$, which is the y-axis. Further examples will show how a curve looks near an asymptote.

HORIZONTAL ASYMPTOTES

For horizontal asymptotes, we are looking to find an equation of a horizontal line as x approaches positive and negative infinity. To find the equation(s):

Let $y = \lim\limits_{x \to +\infty} f(x)$ and $y = \lim\limits_{x \to -\infty} f(x)$

For $f(x)=\frac{1}{x}$ we would get $y=0$ for both of these limits and, as such, our horizontal asymptotes.

Example 6

Determine the vertical and horizontal asymptotes and graph the curve with equation $y = \dfrac{x+1}{x-2}$

For vertical asymptote,

let $x - 2 = 0$

$\qquad x = 2$

For horizontal asymptote(s),

$$y = \lim_{x \to -\infty} \frac{x+1}{x-2} \qquad\qquad y = \lim_{x \to \infty} \frac{x+1}{x-2}$$

$$= \lim_{x \to -\infty} \frac{x+1}{x-2} \times \frac{\frac{1}{x}}{\frac{1}{x}} \qquad = \lim_{x \to \infty} \frac{x+1}{x-2} \times \frac{\frac{1}{x}}{\frac{1}{x}}$$

$$= \lim_{x \to -\infty} \frac{1+\frac{1}{x}}{1-\frac{2}{x}} \qquad\qquad = \lim_{x \to \infty} \frac{1+\frac{1}{x}}{1-\frac{2}{x}}$$

$$= \frac{1+0}{1-0} \qquad\qquad\qquad = \frac{1+0}{1-0}$$

$$y = 1 \qquad\qquad\qquad\qquad y = 1$$

When substituting, for values of x > 2, y >1⇒ above the asymptote and for values of x < 2, y < 1⇒ below the asymptote.

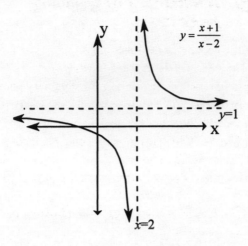

Putting it all together:
A summary of graphing techniques

- State the domain of the function.
- Find the first and second derivatives.
- Perform the first derivative test for stationary points.
- Perform the second derivative test for stationary points.
- State the concavity at each stationary point.
- State whether or not a stationary point is a local maximum, minimum or neither.
- Perform the second derivative test for points of inflection.
- For vertical asymptotes, let the denominator = 0. Give the asymptote(s) equations of the form "$x =$ ".
- For horizontal asymptotes, take the limit as $x \to \pm\infty$. Give the asymptote(s) equations of the form "$y =$ ".
- Use the above information to neatly graph the function.
  - Plot the stationary and inflection points first.
  - Draw the asymptotes with dotted lines.
  - Consider the concavity and nearby x and y values when graphing near the asymptotes.
  - Label your graph fully. Include the equations of the curve and asymptotes, as well as all stationary and inflection points and the x and y-intercepts.

Example 7

Use the graphing techniques of this chapter to graph the function with equation $y = \dfrac{1}{4 - x^2}$

$$D = \{x \in R \mid x \neq \pm 2\}$$

$$y = (4 - x^2)^{-1}$$

$$\frac{dy}{dx} = -1(4 - x^2)^{-2}(-2x)$$

$$= \frac{2x}{(4 - x^2)^2}$$

$$\frac{d^2y}{dx^2} = \frac{2(4 - x^2)^2 - (2x)2(4 - x^2)(-2x)}{(4 - x^2)^4}$$

$$= \frac{6x^2 + 8}{(4 - x^2)^3}$$

For stationary points, let $\dfrac{dy}{dx} = 0$

$$\frac{2x}{(4 - x^2)^2} = 0$$

$$2x = 0 \qquad\qquad \textit{After cross multiplying}$$

$$x = 0$$

$$y = \frac{1}{4}$$

$$\frac{d^2y}{dx^2} = \frac{1}{8} > 0 \quad \therefore \textit{Concave upward}$$

$\therefore$ *Local minimum at* $(0, \dfrac{1}{4})$

For inflection point, let $\dfrac{d^2y}{dx^2} = 0$

$$\frac{6x^2+8}{\left(4-x^2\right)^3}=0$$

$$6x^2+8=0$$

$$x^2=\frac{-8}{6}<0 \quad (undefined)$$

$\therefore$ *no inflection point.*

For vertical asymptotes:

Let $4 - x^2 = 0$
$$x^2 = 4$$
$$x \pm 2$$

For horizontal asymptotes:

$$y = \lim_{x \to \pm\infty} \frac{1}{4 - x^2} \times \frac{\dfrac{1}{x^2}}{\dfrac{1}{x^2}}$$

$$y = \lim_{x \to \pm\infty} \frac{\dfrac{1}{x^2}}{\dfrac{4}{x^2} - 1}$$

$$y = \frac{0}{-1}$$

$$y = 0$$

Notice that the denominator was $-1 < 0$
$\therefore$ *the curve approaches the asymptote from below.*

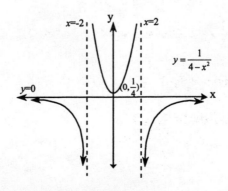

Example 8

Use the graphing techniques of this chapter to graph the function with equation $y = \dfrac{x^2 - 1}{x^2 + 1}$

$$D = \{x \in R\}$$

$$\frac{dy}{dx} = \frac{2x(x^2 + 1) - 2x(x^2 - 1)}{(x^2 + 1)^2}$$

$$= \frac{4x}{(x^2 + 1)^2}$$

$$\frac{d^2y}{dx^2} = \frac{4(x^2 + 1)^2 - (4x)2(x^2 + 1)(2x)}{(x^2 + 1)^4}$$

$$= \frac{4 - 12x^2}{(x^2 + 1)^3}$$

For stationary points, let $\dfrac{dy}{dx} = 0$

$$\frac{4x}{(x^2 + 1)^2} = 0$$

$$4x = 0 \qquad\qquad \textit{After cross multiplying}$$

$$x = 0 \quad y = -1$$

$$\frac{d^2y}{dx^2} = 4 > 0 \quad \therefore \textit{Concave upward}$$

$\therefore$ *Local minimum at* (0,-1).

For inflection points, let $\dfrac{d^2y}{dx^2} = 0$

$$\frac{4 - 12x^2}{(x^2 + 1)^3} = 0$$

$$4 - 12x^2 = 0$$

$$x = \frac{1}{\sqrt{3}} \qquad or \qquad x = -\frac{1}{\sqrt{3}}$$

$$y = -\frac{1}{2} \qquad\qquad\quad y = -\frac{1}{2}$$

Points of inflection at $\left(\dfrac{1}{\sqrt{3}}, -\dfrac{1}{2}\right)$ *and* $\left(-\dfrac{1}{\sqrt{3}}, -\dfrac{1}{2}\right)$

For vertical asymptotes:

Let $x^2 + 1 = 0$

$x^2 = -1$ *undefined*

$\therefore$ *no vertical asymptote.*

For horizontal asymptotes:

$$y = \lim_{x \to \pm\infty} \frac{x^2 - 1}{x^2 + 1} \times \frac{\dfrac{1}{x^2}}{\dfrac{1}{x^2}}$$

$$y = \lim_{x \to \pm\infty} \frac{1 - \dfrac{1}{x^2}}{1 + \dfrac{1}{x^2}}$$

$$y = \frac{1 - 0}{1 + 0}$$

$$y = 1$$

Notice that because we subtract in the numerator and add in the denominator, the numerator is always less than the denominator i.e., $y < 1$. $\therefore$ the curve approaches the asymptote from below.

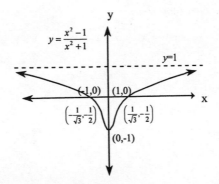

PRACTICE EXERCISE 5

1.

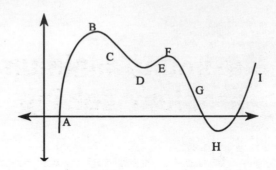

For the diagram above, discuss the concavity for the intervals between the labelled points.

2. Determine the equations of the vertical and horizontal asymptotes for each of the following.

(a) $y = \dfrac{5}{x-1}$ (b) $y = \dfrac{x^2+2}{x^2-9}$

3. Use the graphing techniques of this chapter to graph the function with equation: $y = x^3 - 3x + 2$

4. Use the graphing techniques of this chapter to graph the function with equation:

$$y = \frac{x}{x^2-1}$$

CHAPTER SIX

Maximum/minimum problem solving

In chapter 4, we discussed problem solving with rates. In this chapter, we will be solving problems in which optimal solutions are to be found. The techniques developed in Chapter 5 will be used to help us maximize or minimize quantities.

Steps to follow in max./min. problem solving

- In most problems drawing a diagram helps.
- Develop any necessary equations.
- Define any variables in a space above the equations.
- Decide whether a maximum or minimum solution is required.
- Find the first and second derivatives.
- Apply the first and second derivative tests for stationary points.
- Make sure your solution makes sense. Too often, answers have little relation to the problem or to sensible units of measurement.
- Provide a written conclusion in sentence form, complete with proper units of measurement.

Example 1
DISTANCE AND TIME

An object is thrown into the air. The equation for its height, in metres, after t seconds, is given as $h = -4.9t^2 + 15t + 6$

Find the maximum height and when it is reached.

$$\frac{dh}{dt} = -9.8t + 15$$

$$\frac{d^2h}{dt^2} = -9.8$$

For maximum height, let $\dfrac{dh}{dt} = 0$

$$-9.8t + 15 = 0$$
$$t \doteq 1.53$$
$$h \doteq 17.48$$
$$\frac{d^2h}{dt^2} = -9.8 < 0$$

∴ maximum

A maximum height of 17.48 m occurs after 1.53 s

Example 2
BUSINESS AND ECONOMICS

A pen manufacturer finds that the cost, in dollars, of producing x pens per day is

$$C(x) = 500 + 0.5x + \frac{20\,000}{x}$$

Determine the number of pens that should be produced in order to minimize costs.

$$C'(x) = 0.5 - \frac{20\,000}{x^2}$$

$$C''(x) = \frac{40\,000}{x^3}$$

For minimum, let $C'(x) = 0$

$$0.5 - \frac{20\,000}{x^2} = 0$$

$$x^2 = 40\,000$$

$$x = 200$$

$$C(200) = 700$$

$$C''(200) = 0.005 > 0$$

$\therefore$ *minimum*

A minimum cost of $700 occurs when 200 pens are made.

Hints for developing models for max./min. problem solving

- Picture the problem being described as a situation or event, rather than a math problem. This will help you draw a diagram.
- The value you are looking to maximize or minimize is generally the equation you must develop first.
- Look for clues as to whether a given value is constant or only has its value at a specific moment, and therefore is a variable. Do not substitute its value until after you find the derivatives.
- $ per item might be a clue that the total amount of money needs to be found by multiplying by the number of items.

Hints for developing models for max./min. problem solving (cont'd)

- Perimeter-area-volume problems usually require two formulas to be combined. This usually occurs when given one constant measure and you are asked to maximize or minimize the other. Decide what information is given and how it would simplify the substitution process. (Example 3)
- Distance-time-velocity relationship is frequently manipulated, when, for example, distance variables are written in terms of constant velocity and variable time. (Example 4)
- Relative positioning can result in triangles being drawn. Direction of motion can give clues as to which sides of a triangle are growing, shrinking or remain constant. Look for clues to help decide if right-angled triangles or similar triangles are to be used. (Example 4)

Example 3
SURFACE AREA AND VOLUME

A cylindrical can is to hold 100 mL. Find the dimensions that would minimize the surface area.

Let V represent the capacity of
the can, in mL (or cm³)
Let h be the height, in cm.
Let r be the radius, in cm.
Let S be the surface area, in cm²

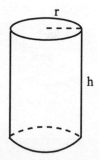

$$V = \pi r^2 h$$
$$S = 2\pi r^2 + 2\pi rh$$

Given $V = 100$ *Volume is constant*

$$\pi r^2 h = 100$$
$$h = \frac{100}{\pi r^2}$$
$$S = 2\pi r^2 + 2\pi r\left(\frac{100}{\pi r^2}\right)$$

Rewrite to substitute for h in the equation for S

$$S = 2\pi r^2 + \frac{200}{r}$$
$$\frac{dS}{dr} = 4\pi r - \frac{200}{r^2}$$
$$\frac{d^2 S}{dr^2} = 4\pi + \frac{400}{r^3}$$

For minimum, let $\dfrac{dS}{dr} = 0$

$$4\pi r - \frac{200}{r^2} = 0$$
$$r^3 = \frac{50}{\pi}$$
$$r = \sqrt[3]{\frac{50}{\pi}} \doteq 2.515$$

$$h = \frac{100}{\pi(2.515)^2} \qquad S = 2\pi(2.515)^2 + \frac{200}{2.515}$$
$$h \doteq 5.035 \qquad\qquad S \doteq 119.265$$
$$\frac{d^2 S}{dr^2} = 4\pi + \frac{400}{(2.515)^3}$$
$$\doteq 37.71$$
$$> 0 \quad \therefore \text{ } minimum$$

A minimum surface area of about 119.3 cm² occurs with radius of 2.5 cm and height of 5.035 cm.

Example 4
DISTANCE, VELOCITY AND TIME

A car leaves an intersection, traveling north at 50 km/h. A truck approaches the same intersection from 20 km away, traveling east at 60 km/h. When will they be closest? How far apart will they be at that time?

Let x be the distance, in km, between the car and the inter-section.
Let y be the distance, in km, between the truck and the intersection.
Let z be the distance, in km, between the car and the truck.
Let t be the time, in hours.

To find when they will be closest, we need to minimize the distance apart.

$x = 50t$

$y = 20 - 60t$

$z^2 = x^2 + y^2$

$z^2 = (50t)^2 + (20 - 60t)^2$

$z^2 = 6100t^2 - 2400t + 400$

Use implicit differentiation on the left side:

$2z\dfrac{dz}{dt} = 12\,200t - 2400$

For minimum, let $\dfrac{dz}{dt} = 0$

$12\,200t - 2400 = 0$

$t \doteq 0.197\ h \quad or \quad 11.8\ min.$

$z^2 = [50(0.197)]^2 + [20 - 60(0.197)]^2$

$z \doteq 12.8$

Use the basic formula:
Distance = velocity x time

50 km/h

z

x

-60 km/h y

*The truck is **approaching** the intersection so the distance to it is decreasing. So its **velocity** can be considered **negative**. The car is **leaving** the intersection, so it has **positive velocity**.*

There is no need to use the second derivative test since we know the truck is approaching faster than the car is leaving ∴ minimum.

The car and truck will be closest after about 0.197 hours, or11.8 minutes.
They will be about 12.8 km apart.

PRACTICE EXERCISE 6

1. A pizza store estimates that the cost per pizza, in dollars, of making x pizzas in a day to be

$$C(x) = 8 + 0.000\,25x + \frac{10}{x}$$

Find the production level that will minimize the cost per pizza.

2. The two equal sides of an isosceles triangle are each 10 cm long. Determine the maximum possible area of this triangle.

3. A boat is sailing north at 12 km/h. Another boat is observed straight ahead at a distance of 15 km, sailing east at 9 km/h. When will the boats be closest?

4. An open-topped box is to be made from 9600 cm² of cardboard. If the length is to be twice the width, determine the dimensions that will maximize the volume.

5. A pipe needs to be laid, connecting point A on one side of a highway to point B, 300 m down the highway and on the other side. The highway is 50 m wide. It costs $150/m to lay the pipe underground and $200/m under the highway. Determine the path to minimize costs.

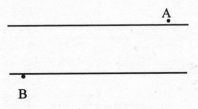

CHAPTER SEVEN

Limits and derivatives of trigonometric functions

In this chapter, limits and derivatives of trigonometric functions and their inverses will be investigated. Please refer to the Coles Notes booklet, *How to Get an A in Trigonometry and Circle Geometry*, to refresh your understanding of trigonometric functions, identities and graphs. In Appendix One of this booklet, we have provided a list of important trigonometric identities that are needed in this chapter.

LIMITS OF TRIGONOMETRIC FUNCTIONS

The following graphs will help in understanding the concepts in this chapter.

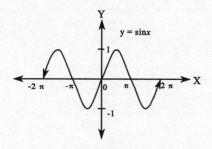

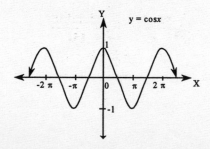

Because they show how trigonometric functions relate to polynomials, the following limits can be developed:

$$\lim_{x \to 0} \frac{\sin x}{x} \quad \text{and} \quad \lim_{x \to 0} \frac{\cos x}{x}$$

To show how they are related, we use a table of values.

x	$\dfrac{\sin x}{x}$	$\dfrac{\cos x}{x}$	x	$\dfrac{\sin x}{x}$	$\dfrac{\cos x}{x}$
0.1	0.9983	9.95	−0.1	0.99833	−9.95
0.01	0.999983	99.995	−0.01	0.9999833	−99.995
0.001	0.9999998	999.9995	−0.001	0.9999998	−999.9995

From the two tables, we can conclude the following:

$$\lim_{x \to 0} \frac{\sin x}{x} = 1 \quad \text{and} \quad \lim_{x \to 0} \frac{\cos x}{x} \text{ is undefined}$$

Example 1

Evaluate each of the following limits by using the identity

$$\lim_{u \to 0} \frac{\sin u}{u} = 1$$

(a) $\lim\limits_{7x \to 0} \dfrac{\sin 7x}{7x}$

$$\lim_{7x \to 0} \frac{\sin 7x}{7x} = 1$$

(b) $\lim\limits_{x \to 0} \dfrac{\sin 15x}{x}$

$$\lim_{x \to 0} \frac{\sin 15x}{x} = \lim_{x \to 0} \frac{15 \sin 15x}{15x}$$
$$= 15 \times 1$$
$$= 15$$

Multiply numerator and denominator by 15

83

(c)

$$\lim_{x \to 0} \frac{\sin x}{3x}$$

$$\lim_{x \to 0} \frac{\sin x}{3x} = \lim_{x \to 0} \frac{1}{3} \times \frac{\sin x}{x} \qquad \textit{Common factor}$$

$$= \frac{1}{3} \times 1$$

$$= \frac{1}{3}$$

Example 2

Use the limits developed above and the basic trigonometric identities (see Appendix One: Important Formulas) to evaluate the following.

(a)

$$\lim_{x \to 0} \frac{\cos^2 x - 1}{x}$$

$$\lim_{x \to 0} \frac{\cos^2 x - 1}{x} = -\lim_{x \to 0} \frac{1 - \cos^2 x}{x}$$

$$= -\lim_{x \to 0} \frac{\sin^2 x}{x} \qquad \mathbf{sin2\theta + cos2\theta = 1}$$

$$= -\lim_{x \to 0} \frac{\sin x}{x} \times \sin x$$

$$= -1 \times 0$$

$$= 0$$

(b)

$$\lim_{x \to 0} \frac{\tan 3x}{x}$$

$$\lim_{x \to 0} \frac{\tan 3x}{x} = \lim_{x \to 0} \frac{\sin 3x}{x \cos 3x} \qquad \mathbf{tan\theta = \frac{sin\theta}{cos\theta}}$$

$$= \lim_{x \to 0} \frac{3 \sin 3x}{3x} \times \frac{1}{\cos 3x}$$

$$= 3 \times 1 \times 1$$

$$= 3$$

84

(c) $\lim\limits_{x \to 0} \dfrac{\cos x - 1}{x}$

$$\lim\limits_{x \to 0} \frac{\cos x - 1}{x} = \lim\limits_{x \to 0} \frac{(\cos x - 1)(\cos x + 1)}{x(\cos x + 1)}$$

To create the Pythagorean relation: $\cos^2 x - 1 = -\sin^2 x$

$$= \lim\limits_{x \to 0} \frac{\cos^2 x - 1}{x(\cos x + 1)}$$

$$= \lim\limits_{x \to 0} \frac{\sin^2 x}{x(\cos x + 1)}$$

$$= \lim\limits_{x \to 0} \frac{\sin x}{x} \times \frac{\sin x}{\cos x + 1}$$

$$= 1 \times \frac{0}{1 + 1}$$

$$= 0$$

Using these techniques, in addition to remembering all the important trigonometric identities, will help evaluate these limits. These concepts are needed for the following section, as well.

DERIVATIVES OF TRIGONOMETRIC FUNCTIONS

Using first principles and the techniques in the preceding section we can develop the derivative of the sine function.

$$y = \sin x$$
$$y + \Delta y = \sin(x + \Delta x)$$
$$\therefore \Delta y = \sin(x + \Delta x) - \sin x$$
$$\frac{dy}{dx} = \lim\limits_{\Delta x \to 0} \frac{\Delta y}{\Delta x}$$

$$= \lim\limits_{\Delta x \to 0} \frac{\sin(x + \Delta x) - \sin x}{\Delta x}$$

Use the trig identity of multiple angles: $\sin(A+B) = \sin A \cos B + \cos A \sin B$

$$= \lim\limits_{\Delta x \to 0} \frac{\sin x \cos \Delta x + \cos x \sin \Delta x - \sin x}{\Delta x}$$

$$= \lim\limits_{\Delta x \to 0} \frac{\sin x(\cos \Delta x - 1) + \cos x \sin \Delta x}{\Delta x}$$

Common factor of $\sin x$

$$= \lim\limits_{\Delta x \to 0} \sin x \left(\frac{\cos \Delta x - 1}{\Delta x} \right) + \lim\limits_{\Delta x \to 0} \cos x \left(\frac{\sin \Delta x}{\Delta x} \right)$$

$$= (\sin x)(0) + (\cos x)(1)$$

See Example 2(c)

$$= \cos x$$

$\therefore$ The derivative of $\sin x$ is $\cos x$.

The following table of trigonometric derivatives is provided without proofs. They are available in all standard calculus textbooks.

FUNCTION	DERIVATIVE
$\sin x$	$\cos x$
$\cos x$	$-\sin x$
$\tan x$	$\sec^2 x$
$\sec x$	$\sec x \tan x$
$\csc x$	$-\csc x \cot x$
$\cot x$	$-\csc^2 x$

Example 3
Find the derivative of each of the following functions.

(a) $y = \sin 3x$

$$\frac{dy}{dx} = \cos 3x \frac{d(3x)}{dx} \qquad \textit{Use the Chain Rule}$$

$$= 3\cos 3x$$

(b) $y = \sin x \tan x$

$$\frac{dy}{dx} = \cos x \tan x + \sin x \sec^2 x \qquad \textit{Use the Product Rule}$$

$$= \cos x \frac{\sin x}{\cos x} + \sin x \sec^2 x$$

$$= \sin x (1 + \sec^2 x)$$

(c) $y = \cos(x^2)$

$$\frac{dy}{dx} = -\sin(x^2)\frac{d(x^2)}{dx} \qquad \textit{Use the Chain Rule}$$

$$= -2x \sin(x^2)$$

(d) $y = \sin^4(x^2 - 1)$

$$\frac{dy}{dx} = 4\sin^3(x^2 - 1)\frac{d\ \sin(x^2-1)}{dx}$$

Use the
Chain Rule

$$= (2x)4\sin^3(x^2 - 1)\ \cos\ (x^2 - 1)$$

$$= 8x\sin^3(x^2 - 1)\ \cos\ (x^2 - 1)$$

(e) $y = \tan(x^3 + 2)$

$$\frac{dy}{dx} = \sec^2(x^3 + 2)\frac{d(x^3 + 2)}{dx}$$

Use the
Chain Rule

$$= 3x^2\sec^2(x^3 + 2)$$

(f) $y = x^2\csc x$

$$\frac{dy}{dx} = 2x\csc x - x^2\csc x\cot x$$

Use the
Product Rule

Example 4
Use implicit differentiation to differentiate $\tan x + \sin y = 1$.

$$\sec^2 x + \cos y\frac{dy}{dx} = 0$$

*Use Chain Rule on **sin** y term*

$$\cos y\frac{dy}{dx} = -\sec^2 x$$

$$\frac{dy}{dx} = \frac{-\sec^2 x}{\cos y}$$

Example 5
Find an equation of the tangent to $y = \cot^2 x$ when $x = \dfrac{\pi}{2}$.

$$\frac{dy}{dx} = 2\cot x\frac{d(\cot x)}{dx}$$

Use Power Rule and Chain Rule

$$= 2\cot x(-\csc^2 x)$$

$$= -2\cot x\csc^2 x$$

$$At \ x = \frac{\pi}{2},$$

$$y = 0$$

$$\frac{dy}{dx} = -2$$

$$\frac{y-0}{x-\dfrac{\pi}{2}} = -2$$

$$y = -2x + \pi$$

Substitute for x

Point-slope form

$\therefore$ *the equation of the tangent at* $x = \dfrac{\pi}{2}$ *is* $y = -2x + \pi$.

APPLICATIONS OF TRIGONOMETRIC DERIVATIVES

Example 6
HARMONIC MOTION

The equation $y = 5\sqrt{3}\cos 2t + 5\sin 2t$ represents the height of a wave in simple harmonic motion, where t is the time in seconds. Find its amplitude, in cm.

**To find the amplitude, we will need to find
the maximum and minimum values.**

$$\frac{dy}{dx} = -10\sqrt{3}\sin 2t + 10\cos 2t$$

$$\frac{d^2y}{dx^2} = -20\sqrt{3}\cos 2t - 20\sin 2t$$

For stationary values, let $\dfrac{dy}{dx} = 0$

$$-10\sqrt{3}\ \sin 2t + 10\cos 2t = 0$$

$$\sqrt{3}\sin 2t = \cos 2t$$

$$\frac{\sin 2t}{\cos 2t} = \frac{1}{\sqrt{3}}$$

$$\tan 2t = \frac{1}{\sqrt{3}}$$

$$2t = \frac{\pi}{6} \quad or \quad 2t = \frac{7\pi}{6} \quad or \quad 2t = \frac{13\pi}{6} \quad \cdots$$

$$y = 10 \qquad\qquad y = -10 \qquad\qquad y = 10$$

$$\frac{d^2y}{dx^2} = -40 \qquad \frac{d^2y}{dx^2} = 40 \qquad \frac{d^2y}{dx^2} = -40$$

maximum   *minimum*   *maximum*

$$Amplitude = \frac{max - min}{2}$$

$$= \frac{10 - (-10)}{2}$$

$$= 10$$

$\therefore$ *The amplitude is 10 cm.*

Example 7
ROTATING LIGHTSOURCE

A lighthouse is situated 200 m from a straight shoreline. The light rotates clockwise at 2 revolutions per minute. At what speed is the beam of light moving along the shore when the angle between the beam and the nearest point on shore is 30º?

Let θ represent the angle of rotation of the lightbeam from the closest point on shore to its position.

Let x represent the distance traveled, in m, along shore to the closest point from the lighthouse.

Let t represent the time, in seconds.

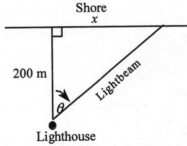

$$\tan\theta = \frac{x}{200}$$

$$\sec^2\theta \frac{d\theta}{dt} = \frac{1}{200}\frac{dx}{dt} \qquad \textbf{\textit{Use Chain Rule for "rate problems"}}$$

$$Given \quad \frac{d\theta}{dt} = 2 \times 360^0$$

$$= 720^0 \qquad\qquad \textbf{\textit{Convert to radian}}$$

$$= 4\pi \; radians \qquad \textbf{\textit{measure because}}$$

$$\qquad\qquad\qquad\qquad\qquad\qquad \textbf{\textit{distance is involved.}}$$

$$\theta = 30^0 = \frac{\pi}{6} \, radians$$

$$\left(\sec^2 \frac{\pi}{6} \right)(4\pi) = \frac{1}{200} \frac{dx}{dt} \qquad \text{\textit{Substitute and solve for} } \frac{dx}{dt}$$

$$\frac{dx}{dt} = \left(\frac{2}{\sqrt{3}} \right)^2 \times 4\pi \times 200$$

$$= 3551.03 m / \min.$$

$$= 55.85 m / s$$

The light is moving along the shore at a speed of 55.85m/s

 ## Tips for solving trigonometric problems

- Properly and fully labelled diagrams are extremely important with trig applications.
- It is generally a good idea to use radian measure, especially when distance measurement is involved.
- Carefully read the problem for clues as to which angle is being used.

INVERSE TRIGONOMETRIC FUNCTIONS

The function $x = \sin y$ is normally written as $y = \sin^{-1} x$, or less commonly as $y = \arcsin x$.

Refer, again, to the Coles Notes booklet, ***How to Get an A in Trigonometry and Circle Geometry***, for further topic development.

The following table gives the derivatives of the three primary inverse trigonometric functions.

FUNCTION	DERIVATIVE
$\sin^{-1} x$	$\dfrac{1}{\sqrt{1-x^2}}$
$\cos^{-1} x$	$\dfrac{-1}{\sqrt{1-x^2}}$
$\tan^{-1} x$	$\dfrac{1}{1+x^2}$

Example 8

Find the derivative of each of the following.

(a) $y = \sin^{-1}\left(4x^3\right)$

$$\frac{dy}{dx} = \frac{1}{\sqrt{1-\left(4x^3\right)^2}} \frac{d\left(4x^3\right)}{dx}$$

*Use the
Chain Rule*

$$= \frac{12x^2}{\sqrt{1-16x^6}}$$

(b) $y = \sqrt{x}\,\tan^{-1} x$

$$\frac{dy}{dx} = \frac{1}{2\sqrt{x}}\tan^{-1} x + \sqrt{x}\,\frac{1}{1+x^2}$$

*Use the
Product Rule*

$$= \frac{\tan^{-1} x}{2\sqrt{x}} + \frac{\sqrt{x}}{1+x^2}$$

Example 9
AVIATION

An airplane is flying at an altitude of 2 km. It is approaching a radar station at a speed of 200 km/h. At what rate is the angle of elevation increasing when the angle of elevation is 60°?

This problem could be solved using standard trigonometric functions, but we will show a solution using inverse functions.

Let x represent the horizontal distance to the radar station

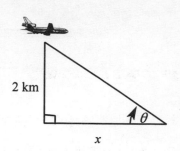

2 km

θ

x

Let θ represent the angle of elevation.

Let t represent the time in h.

Given $\dfrac{dx}{dt} = -200$

Find $\dfrac{d\theta}{dt}$ *when* $\theta = 60°$

$\theta = \tan^{-1}\left(\dfrac{2}{x}\right)$

Use inverse trig derivative with Chain Rule

$\dfrac{d\theta}{dt} = \dfrac{1}{1+\left(\dfrac{2}{x}\right)^2} \times \dfrac{d\left(\dfrac{2}{x}\right)}{dt}$

Being a Rate Problem, use Chain Rule again

$\dfrac{d\theta}{dt} = \dfrac{1}{1+\dfrac{4}{x^2}} \times \dfrac{-2}{x^2} \times \dfrac{dx}{dt}$

$= \dfrac{-2}{x^2+4} \times \dfrac{dx}{dt}$

When $\theta = 60°$, $\tan^{-1}\left(\dfrac{2}{x}\right) = 60°$

Substitute and solve for x

$\dfrac{2}{x} = \sqrt{3}$

$x = \dfrac{2}{\sqrt{3}}$

Substitute $\dfrac{dx}{dt} = -200$ and $\theta = 60°$

$$\dfrac{d\theta}{dt} = \dfrac{-2}{\left(\dfrac{2}{\sqrt{3}}\right)^2 + 4}(-200)$$

$= 75$ *radians / hour*

Substitute and solve for

$$\dfrac{d\theta}{dt}$$

*Whenever distance is used,
radian measure must be used.*

$\therefore$ *the angle of elevation is increasing at 75 rad/h.*

PRACTICE EXERCISE 7

1. Evaluate each of the following limits.

    (a) $\displaystyle\lim_{x\to 0}\dfrac{\sin x}{2x}$

    (b) $\displaystyle\lim_{x\to 0}\dfrac{\sin^2 x}{x}$

    (c) $\displaystyle\lim_{x\to 0}\dfrac{1 - \cos x}{\sin x}$

    (d) $\displaystyle\lim_{x\to 0}\left(x^2 + x^2 \cot^2 x\right)$

2. Find $\dfrac{dy}{dx}$ for each of the following.

    (a) $y = \sin(2x)$

    (b) $y = \cos^3 x$

    (c) $y = \sec\left(x^3\right)$

    (d) $y = \tan(5x - 3)$

    (e) $y = \sin^2 x \cos\left(\dfrac{x}{2}\right)$

    (f) $y = \dfrac{\cos x}{\sec x - \tan x}$

    (g) $\sin x = \tan y$

    (h) $x = \csc^2\left(4y^2\right)$

    (i) $y = \cos^{-1}(5x^2)$

    (j) $y = \left(\tan^{-1} x\right)^2$

3. Find an equation of the tangent to $y = 2 + \cos 2x$ at $x = \dfrac{5\pi}{6}$.

4. Find the local maximum and minimum points and points of inflection for the function:

$$y = \sin^2 x - \frac{x}{2} \quad \text{for} \quad 0 \le x \le \pi$$

Graph the curve.

5. A 6 m ladder is leaning against a wall and begins to slide. The foot of the ladder slides outward at a rate of 0.2 m/s. At what rate is the angle between the ladder and the wall changing when the top is 3 m from the ground?

6. The hypotenuse of a right triangle has length of 10 cm. Determine its maximum possible area.

CHAPTER EIGHT

Exponential and logarithmic functions

This chapter focuses on limits and derivatives of exponential and logarithmic functions. The following graphs will assist you in this chapter. Their domains are stated at the bottom of the graphs.

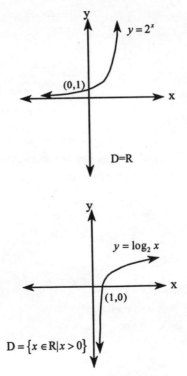

For further help in understanding exponential and logarithmic functions, refer to the Coles Notes, *How to Get an A in Senior Algebra*.

Example 1

Determine each of the following limits.

(a) $\lim\limits_{x \to 0} 5^x$

$$\lim\limits_{x \to 0} 5^x = 5^0$$
$$= 1$$

(b) $\lim\limits_{x \to \infty} 5^x$

$$\lim\limits_{x \to \infty} 5^x = 5^\infty$$
$$= \infty$$

$$\therefore \textbf{ \textit{the limit is undefined}}$$

(c) $\lim\limits_{x \to \infty} 2^{-x}$

$$\lim\limits_{x \to \infty} 2^{-x} = \left(\frac{1}{2} \right)^\infty$$
$$= 0$$

EULER'S NUMBER, e

The number developed below is known as Euler's Number. Developed by Leonhard Euler, a Swiss mathematician, during the 1700's. It is a very important irrational number in Calculus, as we will see in this chapter.

$$e = \lim\limits_{x \to 0} (1 + x)^{\frac{1}{x}}$$

By substituting, we would get the base approaching (but not equal to) 1 and the exponent approaching infinity.

Let's use a table of values.

x	$1 + x$	$e = \lim(1+x)^{\frac{1}{x}}$
100	101	1.047232746
10	11	1.270981615
1	2	2
0.1	1.1	2.5937425
0.01	1.01	2.7048138
0.001	1.001	2.7169239
0.0001	1.0001	2.7181459
0.00001	1.00001	2.7182682
0.000001	1.000001	2.7182805
0.0000001	1.0000001	2.7182817
0.00000001	1.00000001	2.7182818

$$e = \lim_{x \to 0}(1+x)^{\frac{1}{x}}$$
$$= 2.71828182845904523536\ldots$$

which can also be written as: $e = \lim_{x \to \infty}\left(1 + \frac{1}{x}\right)^{x}$

Example 2

Evaluate each of the following limits.

(a)

$$\lim_{x \to 0}(1 + 3x)^{\frac{1}{x}}$$

$$\lim_{x \to 0}(1 + 3x)^{\frac{1}{x}} = \lim_{k \to 0}\left(1 + 3\left(\frac{k}{3}\right)\right)^{\frac{3}{k}} \qquad \textit{Substitute } \frac{k}{3}$$
$$\textit{for } x$$
$$= \lim_{k \to 0}(1 + k)^{\frac{3}{k}}$$
$$= e^{k}$$

97

(b)

$$\lim_{x \to 0}(1+hx)^{\frac{1}{x}}$$

$$\lim_{x \to 0}(1+hx)^{\frac{1}{x}} = e^h \qquad\qquad \textit{See part (a) above}$$

(c)

$$\lim_{x \to 0}(1-x)^{\frac{1}{x}}$$

$$\lim_{x \to 0}(1-x)^{\frac{1}{x}} = e^{-1}$$

NATURAL LOGARITHMS

Recall the definition of a logarithm:

$$\text{If } y = a^x \text{ then } x = \log_a y, \; y > 0$$

and the following properties of logarithms:

1. $\log_a x + \log_a y = \log_a(xy)$

2. $\log_a x - \log_a y = \log_a\left(\dfrac{x}{y}\right)$

3. $\log_a x^b = b \log_a x$

4. $\log_a a = 1$

The **natural logarithm** can now be defined in the following way:

$$\text{If } x = e^y \text{ then } y = \log_e x$$
$$\text{or } y = \ln x, \; x > 0$$

(Pronounced, "lon x")

Properties of e^x and $\ln x$:

1. $\ln e = 1$
2. $\ln e^x = x$
3. $e^{\ln x} = x$
4. $\ln 1 = 0$

Example 3

Solve for x in each of the following.

(a) $\ln x = 12$

$$\log_e x = 12$$
$$x = e^{12} \qquad \textit{Definition of } \ln x$$

(b) $e^{3x+1} = 9$

$$3x + 1 = \ln 9$$
$$x = \frac{\ln 9 - 1}{3} \qquad \textit{Rewrite in logarithmic form}$$

Example 4

On the same set of axes, graph $y = \log_2 x$ and $y = \ln x$.

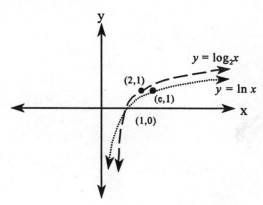

The domain for each function is $D = \{x \in R \mid x > 0\}$

THE DERIVATIVE OF $y = \ln x$

Using first principles, we will find the derivative of $y = \ln x$.

$$\frac{dy}{dx} = \lim_{\Delta x \to 0} \frac{\ln(x + \Delta x) - \ln x}{\Delta x}$$

$$= \lim_{\Delta x \to 0} \frac{\ln\left(\dfrac{x + \Delta x}{x}\right)}{\Delta x} \qquad \textit{Property of logarithms}$$

$$= \lim_{\Delta x \to 0} \frac{1}{\Delta x} \ln\left(1 + \frac{\Delta x}{x}\right) \qquad \textit{Factor } \frac{1}{\Delta x}$$

$$= \lim_{\Delta x \to 0} \frac{x}{x \Delta x} \ln\left(1 + \frac{\Delta x}{x}\right) \qquad \begin{array}{l}\textit{Multiply numerator and} \\ \textit{denominator by } x\end{array}$$

$$= \lim_{\Delta x \to 0} \frac{1}{x} \ln\left(1 + \frac{\Delta x}{x}\right)^{\frac{x}{\Delta x}} \qquad \textit{Property of logarithms}$$

$$= \frac{1}{x} \ln\left[\lim_{\Delta x \to 0}\left(1 + \frac{\Delta x}{x}\right)^{\frac{x}{\Delta x}} \right] \qquad \begin{array}{l}\textit{Take the limit inside the ln} \\ \textit{function}\end{array}$$

$$= \frac{1}{x} \ln e \qquad \textit{Definition of e}$$

$$= \frac{1}{x}(1)$$

$$= \frac{1}{x}$$

If $y = \ln x$

$$\frac{dy}{dx} = \frac{1}{x}$$

Using the chain rule, if $y = \ln u$

$$\frac{dy}{dx} = \frac{1}{u} \cdot \frac{du}{dx}$$

Example 5

Find the derivative of each of the following.

(a) $y = \ln 5x$

$$\frac{dy}{dx} = \frac{1}{5x} \cdot \frac{d(5x)}{dx}$$ Use the chain rule

$$= \frac{1}{5x} \cdot 5$$

$$= \frac{1}{x}$$

(b) $y = \ln\left(x^5\right)$

$$\frac{dy}{dx} = \frac{1}{x^5} \cdot \left(5x^4\right)$$ Use the chain rule

$$= \frac{5}{x}$$

(c) $y = \left(\ln x\right)^3$

$$\frac{dy}{dx} = 3\left(\ln x\right)^2 \left(\frac{1}{x}\right)$$ Use the chain rule
on the exponent first

$$= \frac{3}{x}\left(\ln x\right)^2$$

DERIVATIVE OF $y = e^x$

If $y = e^x$

then $x = \ln y$ *Rewrite in logarithmic form*

$$\frac{dx}{dx} = \frac{1}{y} \cdot \frac{dy}{dx}$$ *Implicit differentiation*

$$1 = \frac{1}{y} \cdot \frac{dy}{dx}$$

$$\frac{dy}{dx} = y$$ *Now substitute y=e^x*

$$\therefore \frac{dy}{dx} = e^x$$

$$\text{If } y = e^x$$

$$\text{then } \frac{dy}{dx} = e^x$$

Using the chain rule, if $y = e^u$

$$\frac{dy}{dx} = e^u \cdot \frac{du}{dx}$$

Example 6

Find the derivative of each of the following.

(a) $y = e^{4x}$

$$\frac{dy}{dx} = e^{4x} \cdot \frac{d(4x)}{dx}$$ *Use the chain rule*

$$= 4e^{4x}$$

(b) $y = e^{x^4}$

$$\frac{dy}{dx} = e^{x^4} \cdot \frac{d(x^4)}{dx}$$ *Use the chain rule*

$$= 4x^3 e^{x^4}$$

(c) $y = e^{\cos x}$

$$\frac{dy}{dx} = e^{\cos x} \cdot \frac{d \cos x}{dx}$$

Use the chain rule and trig derivative

$$= -\sin x \, e^{\cos x}$$

DERIVATIVES OF OTHER EXPONENTIAL AND LOGARITHMIC FUNCTIONS

Without proof, the following important derivatives are provided:

If $y = a^x$

then $\dfrac{dy}{dx} = a^x \ln a$

If $y = \log_a x$

then $\dfrac{dy}{dx} = \dfrac{1}{x \ln a}$

Example 7

Find the derivative of each of the following.

(a) $y = 6^x$

$$\frac{dy}{dx} = 6^x \ln 6$$

(b) $y = 2^{x^2-1}$

$$\frac{dy}{dx} = 2^{x^2-1} \ln 2 \frac{d(x^2-1)}{dx} \qquad \textit{Use the chain rule}$$

$$= \ln 2 \left(2^{x^2-1}\right)(2x)$$

(c) $y = \log_5 x$

$$\frac{dy}{dx} = \frac{1}{x \ln 5}$$

(c) $y = \log_{10}\left(7 - x^2\right)$

$$\frac{dy}{dx} = \frac{1}{\left(7 - x^2\right) \ln 10} \cdot \frac{d\left(7 - x^2\right)}{dx} \qquad \textit{Use the chain rule}$$

$$= \frac{-2x}{\left(7 - x^2\right) \ln 10}$$

Example 8

Use logarithms to find the derivative of $y = x^x$

$$\ln y = \ln x^x$$ *Take* **ln** *of both sides*

$$\ln y = x \ln x$$

$$\frac{1}{y}\frac{dy}{dx} = 1 \cdot \ln x + x \cdot \frac{1}{x}$$ *Implicit on the left and product rule on the right*

$$\frac{1}{x^x}\frac{dy}{dx} = 1 + \ln x$$ *Substitute* $y = x^x$

$$\frac{dy}{dx} = (1 + \ln x)x^x$$

These examples show the importance of the exponential and logarithmic functions.
We now have a function whose **derivative is itself** ($y = e^x$) and one whose **derivative is a reciprocal** ($y = \ln x$).

Further, we can now find the derivative of any exponential or logarithmic function. The following section will provide concrete applications of these derivatives.

APPLICATIONS TO GRAPHING

Example 9

Find all stationary and inflection points and asymptotes and sketch the graph of $y = x^2 e^x$

$$D = R$$ *No restriction on the domain*

$$\frac{dy}{dx} = 2xe^x + x^2 e^x$$ *Use the product rule*

$$= (2x + x^2)e^x$$

$$\frac{d^2 y}{dx^2} = (2 + 2x)e^x + (2x + x^2)e^x$$ *Use the product rule*

$$= (x^2 + 4x + 2)e^x$$

For stationary points, let $\dfrac{dy}{dx} = 0$

$\left(2x + x^2\right)e^x = 0$

$e^x = 0$ *or* $2x + x^2 = 0$

undefined $x(2 + x) = 0$

$\qquad\qquad\quad x = 0$ *or* $x = -2$

$\qquad\qquad\quad y = 0 \qquad\qquad y = 4e^{-2}$

$\qquad\qquad\qquad\qquad\qquad\qquad \doteq 0.5$

$\qquad\quad \dfrac{d^2y}{dx^2} = 2 > 0 \qquad \dfrac{d^2y}{dx^2} = -2e^{-2} < 0$

$\qquad\quad \therefore\ Minimum \qquad \therefore\ Maximum$

For inflection points, let $\dfrac{d^2y}{dx^2} = 0$

$\left(x^2 + 4x + 2\right)e^x = 0$

$x^2 + 4x + 2 = 0$

$x = -2 + \sqrt{2}$ *or* $x = -2 - \sqrt{2}$

$\doteq -0.6 \qquad\qquad\qquad\quad \doteq -3.4$

$y \doteq 0.2 \qquad\qquad\qquad\quad y \doteq 0.4$

For horizontal asymptote:

$y = \lim\limits_{x \to +\infty} x^2 e^x$.   *and* $y = \lim\limits_{x \to -\infty} x^2 e^x$ *Use a table of values*

$\quad = \infty\ \therefore\ undefined \qquad\quad y = 0$

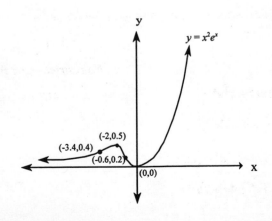

When the **rate of change** of the amount of a substance is **proportional to the amount**, we can write this relation as follows:

$$A(x) \, \alpha \, \frac{dA(x)}{dx}$$

$$\frac{dA(x)}{dx} = k \, A(x), \quad \text{where } k \text{ is a constant}$$

This holds true if $A(x)$ is an exponential function (e^x or a^x)

If $A(x) = ce^{kx}$, where c and k are constants,

then $\dfrac{dA(x)}{dx} = kce^{kx}$

This information will be useful in the following example.

EXPONENTIAL GROWTH

Example 10
DECAY AND HALF-LIFE

Radium decays at a rate that is proportional to its mass. Given that the half-life of radium is 1590 years, and that 20 g of radium was present initially, how many years would it take for 90% of its mass to disintegrate?

Let M represent the mass, in g, after t years

Given $\dfrac{dM}{dt} \, \alpha \, M$

$$\frac{dM}{dt} = kM$$

$$\therefore M = ce^{kt}$$

At $t = 0$, $m = 20$

$$20 = ce^0$$

$$c = 20$$

$$M = 20e^{kt}$$

At $t = 1590,$ $M = 20 \times \dfrac{1}{2} = 10$ *because half-life results*
in half the mass

$10 = 20e^{k(1590)}$

$0.5 = e^{1590k}$

$1590k = \ln 0.5$

$k = \dfrac{\ln 0.5}{1590}$

$M = 20e^{\frac{\ln 0.5}{1590}t}$

90% disintigrates $\Rightarrow$ *10% remains*

$M = 0.1 \times 20 = 2$

$2 = 20e^{\frac{\ln 0.5}{1590}t}$

$\dfrac{\ln 0.5}{1590}t = \ln 0.1$

$t = 5281.87$

It will take about 5282 years for the radium to disintegrate
to 10% of its original mass.

PRACTICE EXERCISE 8

1. Find the derivative of each of the following.

   (a) $y = e^{3x}$

   (b) $y = \ln(x - 8)$

   (c) $y = 2xe^{x}$

   (d) $y = e^{x^3} \ln x$

   (e) $y = \ln\left(x^2 - 5\right)$

   (f) $y = (x - \ln x)e^{x^2}$

   (g) $y = \tan\left(e^{2x}\right)$

   (h) $e^{2x} = \ln y$

   (i) $y = 2^{x^4}$

   (j) $y = \log_{10}(\sec x)$

   (k) $y = \dfrac{e^{5x}}{\ln(5x)}$

   (l) $y = \dfrac{\log_2(x + 1)}{\log_2(x - 1)}$

2. The population of Mathtown increased from 50 000 to 75 000 in 15 years at a rate proportional to the population. Assuming the same rate of growth, determine the size of the town's population after a further 25 years.

3. Given the function $y = (\ln x)^2$
   State the domain of the function;
   Determine the local maximum, minimum and inflection points;
   Determine equations of the vertical and horizontal asymptotes;
   Sketch the curve.

CHAPTER NINE

Integration

This chapter will develop the concept of an integral, first as an anti-derivative, then through applications such as area under a curve and acceleration-velocity-distance.

ANTI-DERIVATIVES

An anti-derivative is the opposite of a derivative. It steps backward to find the function that provides the given derivative.

To find the anti-derivative for $\dfrac{dy}{dx} = 5x^2$, we simply increase the exponent by 1.

Then, instead of multiplying by the exponent, we now divide.

$$y = \frac{5}{3}x^3 + C, \quad C \text{ is a constant}$$

Because the derivative of any constant is 0, we need to add a constant when finding the anti-derivative.

POWER RULE

If $f'(x) = ax^n, \ n \neq -1$

then $f(x) = \dfrac{ax^{n+1}}{n+1} + C$

Example 1

Find the anti-derivative of each of the following.

(a) $f'(x) = x^4 + x^2 + 7x + 5$

$$f(x) = \frac{1}{5}x^5 + \frac{1}{3}x^3 + \frac{7}{2}x^2 + 5x + C$$

Use the power rule

(b) $\dfrac{dy}{dx} = \sqrt{x}$

$$\frac{dy}{dx} = x^{\frac{1}{2}}$$

$$y = \frac{2}{3}x^{\frac{3}{2}} + C$$

Using fractional exponents does not change the technique

INTEGRATION

An anti-derivative is also known as an integral.
The following notation is used:

$$\int f(x)dx = F(x) + C$$

where $F(x)$ is the anti-derivative of $f(x)$.
dx is known as the **differential** and
$f(x)$ is known as the **integrand**.

A list of important standard integrals is provided in Appendix One. They will be used in the following examples.

Example 2

Determine the following integrals.

(a) $\displaystyle\int \cos x \, dx$

$$\int \cos x \, dx = \sin x + C$$

Derivative of sin x is cos x

(b) $\displaystyle\int \frac{1}{x} \, dx$

$$\int \frac{1}{x} \, dx = \ln|x| + C$$

Use absolute value of x because the domain of a logarithmic function must be positive.

110

(c) $\int e^x \, dx$

$$\int e^x \, dx = e^x + C$$

*No change to an
exponential function.*

(d) $\int \left(3e^x - \sec x \tan x \right) dx$

$$\int \left(3e^x - \sec x \tan x \right) dx = 3e^x - \sec x + C$$

The fundamental theorem of Calculus

$$\int_a^b f(x) \, dx = F(b) - F(a)$$

where $F(x)$ is the anti-derivative of $f(x)$.
*This integral is known as a **definite integral**.*

Example 3

Evaluate each of the following integrals.

(a) $\int_{-2}^{5} 6x \, dx$

$$\int_{-2}^{5} 6x \, dx = 3x^2 \Big|_{-2}^{5}$$

$$= 3(5)^2 - 3(-2)^2$$

$$= 63$$

*Find the anti-deriva-
tive, then substitute
and subtract.*

(b) $\int_{0}^{\pi} \sin x \, dx$

$$\int_{0}^{\pi} \sin x \, dx = -\cos x \Big|_{0}^{\pi}$$

$$= -\cos \pi - (-\cos 0)$$

$$= 2$$

*Anti-derivative
of sin x
is −cos x*

111

(c) $\displaystyle\int_{-1}^{2}\left(2e^x - \frac{5}{x}\right)dx$

$$\int_{-1}^{2}\left(2e^x - \frac{5}{x}\right)dx = \left[2e^x - 5\ln|x|\right]_{-1}^{2}$$
$$= \left[2e^2 - 5\ln|2|\right] - \left[2e^{-1} - 5\ln|-1|\right]$$
$$= 2e^2 - 2e^{-1} - 5\ln 2 + 5\ln 1$$
$$= 2e^2 - 2e^{-1} - 5\ln 2 \qquad \qquad ln\ 1 = 0$$

METHODS OF INTEGRATION: SUBSTITUTION

Problems such as $\displaystyle\int 2x\sqrt{1-x^2}\,dx$ and $\displaystyle\int x^2 e^{3x^4}$ can be very difficult.

However, careful analysis will show that the inverse of the chain rule is present in the integrand, which will make the integral easier to simplify.

In the first integral, if we let $u=1-x^2$ and then $du=-2x\,dx$, we can rewrite the integral as :

$$-\int \sqrt{u}\,du$$

which is definitely much simpler to work with. We will complete this as an example later. This principle, however, leads us to what is known as the **Substitution Rule**.

The substitution rule

If we have an integral $\displaystyle\int f(x)\,dx = \int f(g(x))g'(x)\,dx$

then we can let $g(x) = u$ and $g'(x)dx = du$

and rewrite the integral as:

$$\int f(u)\,du$$

Example 4

Determine each of the following integrals.

(a) $\int 2x\sqrt{1-x^2}\, dx$

Substitute for the expression $1 - x^2$ and for $2x\, dx$

$$Let\ \ u = 1 - x^2$$

$$du = -2x\, dx$$

$$\int 2x\sqrt{1-x^2}\, dx = -\int \sqrt{u}\, du$$

$$= -\int u^{\frac{1}{2}}\, du$$

$$= -\frac{2}{3}u^{\frac{3}{2}} + C$$

$$= -\frac{2}{3}\left(1-x^2\right)^{\frac{3}{2}} + C$$

(b) $\int x^3 e^{3x^4}\, dx$

$$Let\ \ u = 3x^4$$

$$du = 12x^3\, dx$$

$$x^3\, dx = \frac{1}{12}\, du$$

Re-arrange so that correct substitution can be made

$$\int x^3 e^{3x^4}\, dx = \int \frac{1}{12}e^u\, du$$

$$= \frac{1}{12}e^u + C$$

$$= \frac{1}{12}e^{3x^4} + C$$

Knowing how the anti-derivative of e^u works, we substitute for the exponent.

Be careful with the substitution method. It only works when the **derivative of u** is present in the integrand. Other helpful methods are shown in the following pages.

(c) $\int \sin^3 x \cos x \, dx$

> *Let* $u = \sin x$
>
> $du = \cos x \, dx$
>
> $\int \sin^3 x \cos x \, dx = \int u^3 \, du$
>
> $= \dfrac{1}{4} u^4 + C$
>
> $= \dfrac{1}{4} \sin^4 x + C$

A complex trig expression becomes an easy algebraic expression.

(d) $\int \dfrac{x}{5 - 2x^2} \, dx$

> *Let* $u = 5 - 2x^2$
>
> $du = -4x \, dx$
>
> $x \, dx = -\dfrac{1}{4} du$
>
> $\int \dfrac{x}{5 - 2x^2} \, dx = -\int \dfrac{\frac{1}{4}}{u} \, du$
>
> $= -\dfrac{1}{4} \ln|u| + C$
>
> $= -\dfrac{1}{4} \ln\left|5 - 2x^2\right| + C$

The derivative of the denominator exists in the numerator. This allows us to use the logarithmic integral.

METHODS OF INTEGRATION: BY PARTS

Frequently, two unrelated functions are together in an integral and the substitution method cannot be used. For example, an algebraic and a trig function together:

$$\int x \cos x \, dx$$

The following method would break the integrand into 2 parts, one of which would be placed with the differential. The integral would then take the form:

$$\int u \, dv$$

To eventually get to this form, start with a function $y = u \cdot v$.

Begin by taking the derivative using the product rule:

$$\frac{d(uv)}{dx} = v\frac{du}{dx} + u\frac{dv}{dx}$$

or:

$$d(uv) = v\,du + u\,dv$$

Then integrate:

$$\int d(uv) = \int v\,du + \int u\,dv$$

$$uv = \int v\,du + \int u\,dv$$

Rewrite:

$$\int u\,dv = uv - \int v\,du$$

INTEGRATION BY PARTS

If an integral can be written in the form:

$$\int u\,dv$$

then the integral can be rewritten as:

$$\int u\,dv = uv - \int v\,du$$

Example 5

Determine each of the following integrals.

(a) $\int x\cos x\,dx$

Let $u = x$ and $dv = \cos x\,dx$

$\therefore du = dx$ and $v = \sin x$

$$\int x\cos x\,dx = uv - \int v\,du$$

$$= x\sin x - \int \sin x\,dx$$

$$= x\sin x + \cos x + C$$

We let $u = x$ because, in the differential, the x term disappears, whereas cos x simply changes to sin x.

(b) $\int \ln x \, dx$

$$\text{Let } u = \ln x \text{ and } dv = dx$$

$$\therefore du = \frac{1}{x} dx \text{ and } v = x$$

$$\int \ln x \, dx = uv - \int v \, du$$

$$= x \ln x - \int dx$$

$$= x \ln x - x + C$$

We let u = ln x because the differential changes to a simple algebraic expression.

(c) $\int x^2 e^x \, dx$

$$\text{Let } u = x^2 \text{ and } dv = e^x \, dx$$

$$\therefore du = 2x \, dx \text{ and } v = e^x$$

$$\int x^2 e^x \, dx = uv - \int v \, du$$

$$= x^2 e^x - 2 \int x e^x \, dx$$

$$\text{Let } u = x \text{ and } dv = e^x \, dx$$

$$\therefore du = dx \text{ and } v = e^x$$

$$\int x^2 e^x \, dx = x^2 e^x - 2 \left[uv - \int v \, du \right]$$

$$= x^2 e^x - 2 \left[x e^x - \int e^x \, dx \right]$$

$$= x^2 e^x - 2x e^x + 2e^x + C$$

We let u=x² because the exponent is reduced in the differential.

The integral has been simplified but we need to integrate by parts a second time.

Hints for integration by parts

If $\int$ (polynomial)(trig) dx,   let u = polynomial

If $\int$(polynomial)(exponential) dx,   let u = polynomial

If $\int$ (polynomial)(logarithmic) dx, let u = logarithmic

If $\int$ (polynomial)(inverse trig) dx, let u = inverse trig

If $\int$ (exponential)(trig) dx,   let u = either.

then apply twice and rewrite.

Example 6

Determine the integral $\int e^x \sin x \, dx$

   Let $u = e^x$ and $dv = \sin x \, dx$

   $\therefore du = e^x \, dx$ and $v = -\cos x$ *Let u = either part.*

   $\int e^x \sin x \, dx = uv - \int v \, du$

   $\qquad\qquad = -e^x \cos x + \int e^x \cos x \, dx$

   Let $u = e^x$ and $dv = \cos x \, dx$ *Repeat the process*

   $\therefore du = e^x \, dx$ and $v = \sin x$

   $\int e^x \sin x \, dx = -e^x \cos x + \left[uv - \int v \, du \right]$

   $\qquad\qquad = -e^x \cos x + \left[e^x \sin x - \int e^x \sin x \, dx \right]$

   $\qquad\qquad = -e^x \cos x + e^x \sin x - \int e^x \sin x \, dx$

The integral on the left side is also found on the right side of the equation. Rewrite and solve for the integral.

$$2\int e^x \sin x \, dx = -e^x \cos x + e^x \sin x$$

$$\therefore \int e^x \sin x \, dx = \frac{1}{2}\left(-e^x \cos x + e^x \sin x\right) + C$$

   Because both exponential and trigonometric functions are periodic in nature (the function repeats or eventually repeats) when differentiating, it allows us to eventually return to the same integral. Then we can rewrite and collect like terms.

METHODS OF INTEGRATION:
BY PARTIAL FRACTIONS

Partial Fractions is a technique in which rational expressions are broken apart into manageable rational expressions, called partial fractions. This is done **if the denominator can be factored**.

Example 7
Use partial fractions to determine the integral $\int \dfrac{x+2}{x^2+4x+3}\,dx$

$\int \dfrac{x+2}{x^2+4x+3}\,dx = \int \dfrac{x+2}{(x+3)(x+1)}\,dx$ *Factor the denominator*

Let $\dfrac{x+2}{(x+3)(x+1)} = \dfrac{A}{x+3} + \dfrac{B}{x+1}$ *Rewrite into partial fractions.*

Multiply both sides by $(x+3)(x+1)$

$x+2 = A(x+1) + B(x+3)$

$x+2 = Ax + A + Bx + 3B$

$(A+B)x + (A+3B) = x+2$

Multiply by the common denominator and reduce

Comparing coefficients,

$\therefore A + B = 1$

$A + 3B = 2$

$\therefore A = \dfrac{1}{2},\ B = \dfrac{1}{2}$

Solve the system of equations to find A and B

$\int \dfrac{x+2}{(x+3)(x+1)}\,dx = \int \dfrac{A}{x+3}\,dx + \int \dfrac{B}{x+1}\,dx$

$= \int \dfrac{\frac{1}{2}}{x+3}\,dx + \int \dfrac{\frac{1}{2}}{x+1}\,dx$

$= \dfrac{1}{2}\ln|x+3| + \dfrac{1}{2}\ln|x+1| + C$

Place the integral into partial so that we may use $\int \dfrac{1}{u}\,du = \ln|u|$

Example 8

Use partial fractions to determine the integral $\int \dfrac{4x+13}{x^2+6x+9}\,dx$

$$\int \frac{4x+13}{x^2+6x+9}\,dx = \int \frac{4x+13}{(x+3)^2}\,dx$$

Factor the denominator

Let $\quad \dfrac{4x+13}{(x+3)^2} = \dfrac{A}{x+3} + \dfrac{B}{(x+3)^2}$

Multiply through by $(x+3)^2$

$$4x+13 = A(x+3)+B$$

$$4x+13 = Ax+(3A+B)$$

$$Ax = 4x$$

$$\therefore A = 4$$

$$3A+B = 13$$

$$\therefore A = 4, \quad B = 1$$

*Rewrite into partial fractions. Note the second denominator. This is because $(x+3)^2$ represents the **greatest** common denominator.*

$$\int \frac{4x+13}{(x+3)^2}\,dx = \int \frac{A}{x+3}\,dx + \int \frac{B}{(x+3)^2}\,dx$$

$$= \int \frac{4}{x+3}\,dx + \int \frac{1}{(x+3)^2}\,dx$$

$$= 4\ln|x+3| - \frac{1}{x+3} + C$$

Use substitution method on the second integral.

METHODS OF INTEGRATION: BY TRIGONOMETRIC SUBSTITUTIONS

Integrals containing powers of trigonometric functions are seemingly difficult. However, using the Pythagorean relation, $\sin^2 x + \cos^2 x = 1$, can sometimes simplify the integral.

Example 9

Determine the integral $\int \sin^3 x\,dx$

$$\int \sin^3 x\,dx = \int \left(\sin^2 x\right)\sin x\,dx$$

Substitute for $\sin^2 x$

$$= \int \left(1-\cos^2 x\right)\sin x\,dx$$

Expand the brackets

$$= \int \sin x\,dx - \int \cos^2 x \sin x\,dx$$

$$= -\cos x + \frac{1}{3}\cos^3 x + C$$

Use substitution method on the second integral.

119

In many examples, the algebraic Pythagorean relation, $\sqrt{a^2 - x^2}$, occurs. In such examples, you may be able to substitute $x = a\cos\theta$ or $x = a\sin\theta$.

Example 10
Determine the integral $\displaystyle\int \frac{dx}{\sqrt{9 - x^2}}$

Let $x = 3\cos\theta$

$\therefore dx = -3\sin\theta\, d\theta$

$\displaystyle\int \frac{dx}{\sqrt{9 - x^2}} = \int \frac{-3\sin\theta}{\sqrt{9 - 9\cos^2\theta}} d\theta$

$\displaystyle = \int \frac{-3\sin\theta}{\sqrt{9\sin^2\theta}} d\theta$

$\displaystyle = \int \frac{-3\sin\theta}{3\sin\theta} d\theta$

$\displaystyle = -\int d\theta$

$= -\theta + C$

$\displaystyle = -\cos^{-1}\left(\frac{x}{3}\right) + C$

The Pythagorean relation allows us to use trig substitutions

*We must now replace θ with the appropriate **inverse** trig function.*

Similarly, if $\sqrt{x^2 - a^2}$ occurs, we may be able to replace x with $a\tan\theta$, and if $\sqrt{x^2 - a^2}$ occurs, we may be able to replace x with $a\sec\theta$.

PRACTICE EXERCISE 9

1. Integrate each of the following.

(a) $\int \left(4x^3 - 5x \right) dx$

(b) $\int \left(e^{2x} + e^{-2x} \right) dx$

(c) $\int \left(\dfrac{1}{x} - \dfrac{2}{x^2} - \dfrac{3}{x^3} \right) dx$

(d) $\int \sin^2 x \cos x \, dx$

(e) $\int_{3}^{13} 2x \sqrt{x^2 - 5} \, dx$

(f) $\int x \sec^2 x \, dx$

(g) $\int e^x (5x + 2) \, dx$

(h) $\int_{1}^{e} (\ln x)^2 \, dx$

(i) $\int x \sqrt{x + 2} \, dx$

(j) $\int \dfrac{11x + 7}{x^2 + 4x - 5} \, dx$

(k) $\int_{3}^{4} \dfrac{2x + 3}{(x - 2)^2} \, dx$

(l) $\int \dfrac{dx}{\sqrt{16 - x^2}}$

(m) $\int_{0}^{\pi} \cos^3 x \, dx$

121

CHAPTER TEN

Applications
of integration

AREA UNDER A CURVE

Given a curve,
$y = f(x)$, as shown to the right,
between $x = a$ and $x = b$. We can
find the area under the curve.

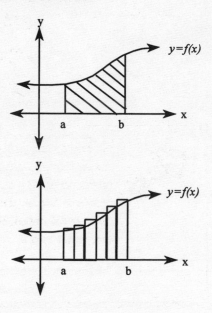

We will do so by dividing
the region up into thin rectangles,
all with width of Δx, as shown in
the second diagram on the right.

If we allow the width of
each rectangle to approach zero,
it would result in a smooth curve.
The third diagram on the right
shows a blow-up of a typical thin
rectangle.

The area of the lower part of the rectangle is
$l \times w = y\Delta x$

The area of the whole large rectangle is
$(y + \Delta y)\Delta x$

The shaded area within the rectangle under the
curve represents the change in area, or ΔA

$\therefore$ We can say that $\quad y\Delta x \leq \Delta A \leq (y + \Delta y)\Delta x$

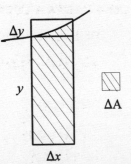

As was mentioned above, if we allow Δx to approach zero, we would get a smooth curve. However, dividing through by Δx will enable us to use some of our calculus methods from chapters 1 and 2.

$$y \le \frac{\Delta A}{\Delta x} \le (y + \Delta y)$$

$$\lim_{\Delta x \to 0} y \le \lim_{\Delta x \to 0} \frac{\Delta A}{\Delta x} \le \lim_{\Delta x \to 0}(y + \Delta y)$$

The first limit becomes y

The second limit is the definition of $\dfrac{dA}{dx}$

In the third limit, as $\Delta x \to 0$, $\Delta y \to 0$

$$y \le \frac{dA}{dx} \le y$$

$$\therefore \frac{dA}{dx} = y$$

$\therefore$ the rate of change of the area with respect
to x is the function $y = f(x)$

The area under the curve is the sum of these thin rectangles. In fact, this is the true definition of integration. By the development above, we can now say that the area under the curve, above the x-axis, is the **anti-derivative**, or integral, of $y = f(x)$ between a and b.

$$A = \int_{a}^{b} f(x)\, dx$$

Example 1
AREA UNDER A CURVE

Find the area under the curve $f(x)=3x^2 + 2$, between x=−2 and x=3.

$$A = \int_{-2}^{3} \left(3x^2 + 2\right) dx$$

$$= \left[x^3 + 2x\right]_{-2}^{3}$$

$$= 33 - (-12)$$

$$= 45 \; units^2$$

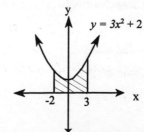

Example 2
AREA BETWEEN TWO CURVES

Determine the area bounded by the curves with equations $y=5x^2-1$ and $y=x^2+8$.

To find the points of intersection,
$$let\ \ 5x^2 - 1 = x^2 + 8$$
$$4x^2 - 9 = 0$$
$$x = \pm 1.5$$

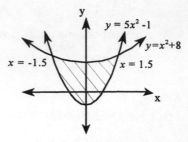

$$A = \left| \int_{-1.5}^{1.5} \left[\left(5x^2 - 1\right) - \left(x^2 + 8\right) \right] dx \right|$$

$$= \left| \int_{-1.5}^{1.5} \left(4x^2 - 9\right) dx \right|$$

$$= \left| \frac{4}{3}x^3 - 9x \right|_{-1.5}^{1.5}$$

$$= \left| -9 - 9 \right|$$

$$= 18\ units^2$$

We take the absolute value because, when subtracting, we may get a negative area that should be positive.

VOLUMES OF REVOLUTION

Circular solids are produced when curves are rotated about the x or y-axis. When finding volumes of revolution, we can use similar techniques to those used in finding areas.

In the diagram on the right, the curve $y = f(x)$ is rotated about the x-axis. When discussing the area under the curve, we used thin rectangles with width of Δx.

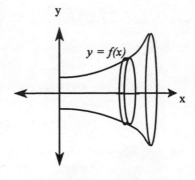

For volumes of revolution, divide the solid into many thin cylinders or discs, each with thickness, or height of Δx and radius of y.

The volume of each disc will be

$$V = \pi r^2 h$$
$$= \pi y^2 \Delta x$$
$$= \pi [f(x)]^2 \Delta x$$

The total volume of revolution about the x-axis, for $a \leq x \leq b$, will then be: $V = \int_a^b \pi [f(x)]^2 \, dx$

Example 3
VOLUME OF REVOLUTION

Find the volume of the solid when the region under the curve $y=e^x$ is rotated about the x-axis, for $-1 \leq x \leq 2$.

$$V = \int_{-1}^2 \pi [e^x]^2 \, dx$$
$$= \int_{-1}^2 \pi e^{2x} \, dx$$
$$= \left[\frac{\pi}{2} e^{2x} \right]_{-1}^2$$
$$= \frac{\pi}{2} \left(e^4 - e^{-2} \right) units^2$$

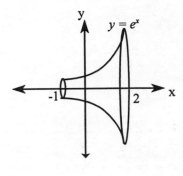

Slope applications

Recall that the slope of a curve can be found by finding the derivative of the equation of the curve. In the following example, the slope is given. To find the equation of the curve, we simply find the anti-derivative.

Example 4
SLOPE AT A POINT

The slope of a curve is given by the equation $\dfrac{dy}{dx} = 6x^2 - 4x + 2$

If the curve passes through the point (1,5), find the equation of the curve.

$$y = \int \left(6x^2 - 4x + 2\right) dx$$
$$y = 2x^3 - 2x^2 + 2x + C$$
$$At\ x = 1,\ y = 5$$
$$5 = 2(1)^3 - 2(1)^2 + 2(1) + C$$
$$C = 3$$
$$y = 2x^3 - 2x^2 + 2x + 3$$

VELOCITY AND ACCELERATION

Recall that the derivative of displacement is velocity and of velocity is acceleration. The following examples turn this around. Integrating allows us to find the height or displacement.

Example 5
HEIGHT VS GRAVITY

A ball is thrown upwards with initial velocity of 15 m/s and initial height of 1.5 m, against constant gravitational acceleration of 9.8 m/s². Find the height after 2 s.

$$a = -9.8$$

*Velocity is the **anti-derivative** of acceleration.*

$$v = \int (-9.8)\, dt$$
$$v = -9.8t + v_0$$
$$v = -9.8t + 15$$

For a polynomial function, the constant, C, can be replaced by v_0, the initial velocity.

$$s = \int (-9.8t + 15)\, dt$$
$$s = -4.9t^2 + 15t + s_0$$
$$s = 4.9t^2 + 15t + 1.5$$
$$s(2) = 4.9(2)^2 + 15(2) + 1.5$$
$$s(2) = 51.1$$

Replace the constant with s_0, the initial height.

After 2 s, the height will be 51.1 m.

Example 6
HARMONIC MOTION

A particle oscillates in simple harmonic motion with velocity given as $v = 24\pi \cos 3\pi t$ cm/s, where t is time, in s. If initial displacement is 0 cm, determine the displacement after 2.5 s.

$$s(t) = \int (24\pi \cos 3\pi t)\, dt$$ *Displacement is the anti-derivative of velocity.*

$$s(t) = \frac{24\pi}{3\pi} \sin 3\pi t + C$$

$$s(0) = 0 \quad \therefore C = 0$$ *Given $s_0 = 0$. Substitute.*

$$\therefore s(t) = 8 \sin 3\pi t$$

$$s(2.5) = 8 \sin 3\pi (2.5)$$

$$s(2.5) = -8$$

After 2.5 s, the displacement is −8 cm.

ECONOMICS

Economics

Just as we did in previous examples, we can reverse the process here. Given the marginal cost, the total cost function would be the anti-derivative of the marginal cost function. It is important to note that **fixed costs occur before production takes place**. We can then substitute $x=0$, given the fixed costs.

Example 7
PRODUCTION COSTS

The Perfect Pen Company has determined the marginal cost of producing x pens to be $\dfrac{5}{\sqrt{x}}$ dollars per pen.

pens. If ~~~~ costs are $2000, find the total cost of producing 10 000 pens.

Let C(x) represent the cost of producing x pens.

$$Given \quad C'(x) = \frac{5}{\sqrt{x}}$$

$$= 5x^{-\frac{1}{2}}$$

$$C(x) = \int 5x^{-\frac{1}{2}} \, dx$$

$$C(x) = 10x^{\frac{1}{2}} + K$$

$$C(0) = 2000 \quad \therefore K = 2000$$

$$C(x) = 10\sqrt{x} + 2000$$

$$C(10\,000) = 10\sqrt{10\,000} + 2000$$

$$C(10\,000) = 3000$$

Integrate the marginal cost to find the total cost function.

Fixed costs occur before production, i.e., at x=0

The total cost of producing 10 000 pens is $3000.

THE LAW OF NATURAL GROWTH

In Chapter 8, exponential growth was investigated. If the rate of change of a quantity is proportional to the quantity present, the function would be expressed as an exponential.

If a population's rate of change is proportional to the population, it is said to have Natural Growth.

$$If \quad \frac{dP}{dt} = kP$$

$$then \quad P = Ce^{kt}$$

$$or \; better \quad P = P_0e^{kt}$$

Example 8
EXPONENTIAL GROWTH

The population of a town grows at a rate of 5% per year. How long will it take for the population to double?

Let P represent the size of the population after t years.

$$\frac{dP}{dt} = kP$$

$$\frac{dP}{dt} = 0.05P \qquad \text{*This follows the Law of*}$$
$$\text{*Natural Growth*}$$

$$P = P_0 e^{0.05t}$$

For the population to double, $P = 2P_0$

$$2P_0 = P_0 e^{0.05t}$$

$$e^{0.05t} = 2 \qquad \text{*Rewrite in logarithmic form*}$$

$$0.05t = \ln 2$$

$$t \doteq 13.86$$

The population should double in about 14 years.

NEWTON'S LAW OF COOLING

Newton's Law of Cooling states that the rate at which the temperature of a body changes is proportional to the temperature difference between the body, T, and that of its surroundings, S.

$$\frac{dT}{dt} = k(T - S)$$

$$T - S = Ce^{kt}$$

C would be the initial temperature difference between the body and the surroundings.

$$C = T_0 - S$$

$$T - S = (T_0 - S)e^{kt}$$

$$\therefore T = S + (T_0 - S)e^{kt}$$

Example 9
LAW OF COOLING

A drink had a temperature of 20°C when it was placed into a refrigerator with internal temperature of 5°C. The temperature of the drink was 17° one hour later. When will the temperature reach 8°?

Let T represent the temperature of the drink after t hours.
Let S represent the internal temperature of the refrigerator.

$$T = S + (T_0 - S)e^{kt}$$ *Newton's Law of Cooling*

$$T = 5 + (20 - 5)e^{kt}$$

$$T = 5 + 15e^{kt}$$

At $t = 1,$ $T = 17$

$$17 = 5 + 15e^{k(1)}$$

$$e^k = 0.8$$

$$k = \ln 0.8$$

$$T = 5 + 15e^{(\ln 0.8)t}$$

Substitute $T = 8$

$$8 = 5 + 15e^{(\ln 0.8)t}$$

$$0.2 = e^{(\ln 0.8)t}$$

$$(\ln 0.8)t = \ln 0.2$$

$$t \doteq 7.2$$

It will take about 7.2 h to cool to 8°C.

PRACTICE EXERCISE 10

1. The slope of a curve is given as $2\sqrt{x}+6$

   The point (2,5) lies on the curve. Find an equation of the curve.

2. A pebble is tossed straight upward at 25 m/s from the edge of a bridge, 46 m above a lake.

   How many seconds later does the pebble hit the lake?

3. The temperature of a cup of coffee is 80°C when brought into a room where the temperature is 22°C. After 10 minutes, the temperature of the coffee is 60°C. How long will it take for the temperature to reach 30°C?

4. Determine the area of the region bounded by the curves with equations $y = x^2 - 2x - 4$ and $y = -x^2 + 8$.

5. Find the volume of the solid obtained when the area under $y = \dfrac{1}{\sqrt{x}}$ is rotated about the x- axis for $2 \leq x \leq 6$.

6. Two alternators are connected so that the rate of change of the current, I, is given as

   $-120\pi\sqrt{3}\cos 120\pi t \sin - 120\pi t$ amperes/hour

   Given that the initial current is 1 amp, determine
   (a) an equation for the current;
   (b) the maximum current.

7. The marginal profit when selling x items is $x\sqrt{x^2+25}$.
   Determine the total profit when selling 50 items.

Important formulas

TRIGONOMETRIC IDENTITIES

$$\sec x = \frac{1}{\cos x} \qquad \csc x = \frac{1}{\sin x}$$

$$\tan x = \frac{\sin x}{\cos x} \qquad \cot x = \frac{1}{\tan x}$$

$$\sin^2 x + \cos^2 x = 1$$

$$\sec^2 x = 1 + \tan^2 x$$

$$\csc^2 x = 1 + \cot^2 x$$

$$\sin(x + y) = \sin x \cos y + \cos x \sin y$$

$$\sin(x - y) = \sin x \cos y - \cos x \sin y$$

$$\cos(x + y) = \cos x \cos y - \sin x \sin y$$

$$\cos(x - y) = \cos x \cos y + \sin x \sin y$$

$$\tan(x + y) = \frac{\tan x + \tan y}{1 - \tan x \tan y}$$

$$\tan(x - y) = \frac{\tan x - \tan y}{1 + \tan x \tan y}$$

$$\sin 2x = 2 \sin x \cos x$$

$$\cos 2x = \cos^2 x - \sin^2 x$$

$$= 2 \cos^2 x - 1$$

$$= 1 - 2 \sin^2 x$$

VOLUME AND SURFACE AREA

CYLINDER	CONE	SPHERE
$V = \pi r^2 h$	$V = \dfrac{1}{3}\pi r^2 h$	$V = \dfrac{4}{3}\pi r^3$
$S = 2\pi r^2 + 2\pi r h$	$S = \pi r^2 + \pi r \sqrt{r^2 + h^2}$	$S = 4\pi r^2$

DERIVATIVE RULES

Power ax^n nax^{n-1}

Sum $f(x) + g(x)$ $f'(x) + g'(x)$

Product $f(x) \cdot g(x)$ $f'(x)g(x) + g'(x)f(x)$

$u \cdot v$ $\dfrac{du}{dx} \cdot v + \dfrac{dv}{dx} \cdot u$

Quotient $\dfrac{f(x)}{g(x)}$ $\dfrac{f'(x)g(x) - g'(x)f(x)}{[g(x)]^2}$

$\dfrac{u}{v}$ $\dfrac{\dfrac{du}{dx} \cdot v - \dfrac{dv}{dx} \cdot u}{v^2}$

Chain $f(g(x))$ $f'(g(x)) \cdot g'(x)$

$f(u(x))$ $\dfrac{dy}{du} \cdot \dfrac{du}{dx}$

DERIVATIVE BY FIRST PRINCIPLES

$$f'(x) = \lim_{\Delta x \to 0} \frac{f(x + \Delta x) - f(x)}{\Delta x}$$
$$= \lim_{h \to 0} \frac{f(x + h) - f(x)}{h}$$

DERIVATIVES OF SPECIFIC FUNCTIONS

$$\frac{d \sin x}{dx} = \cos x \qquad\qquad \frac{de^x}{dx} = e^x$$

$$\frac{d \cos x}{dx} = -\sin x \qquad\qquad \frac{d \ln x}{dx} = \frac{1}{x}$$

$$\frac{d \tan x}{dx} = \sec^2 x \qquad\qquad \frac{da^x}{dx} = a^x \ln x$$

$$\frac{d \sec x}{dx} = \sec x \tan x \qquad\qquad \frac{d \log_a x}{dx} = \frac{1}{x \ln a}$$

$$\frac{d \csc x}{dx} = -\csc x \cot x \qquad\qquad \frac{d \sin^{-1} x}{dx} = \frac{1}{\sqrt{1-x^2}}$$

$$\frac{d \cot x}{dx} = -\csc^2 x \qquad\qquad \frac{d \cos^{-1} x}{dx} = \frac{-1}{\sqrt{1-x^2}}$$

$$\frac{dC}{dx} = 0 \qquad\qquad \frac{d \tan^{-1} x}{dx} = \frac{1}{1+x^2}$$

INTEGRATION TECHNIQUES

SUBSTITUTION

If $\int f(x)\, dx = \int f(g(x))g'(x)\, dx$

then let $u = g(x)$ and rewrite as:

$$\int f(u)\, du$$

PARTS

If $\int f(x)\, dx$ can be written as $\int u\, dv$

then $\int u\, dv = uv - \int v\, du$

PARTIAL FRACTIONS

If in $\int \frac{f(x)}{g(x)}\, dx$, both $f(x)$ and $g(x)$ are polynomials,

and if $g(x)$ is factorable, i.e. $g(x) = (x-p)(x-q)\ldots$

then let $\dfrac{f(x)}{g(x)} = \dfrac{A}{x-p} + \dfrac{B}{x-q} + \ldots$

and solve for A, B, . . . before integrating

INTEGRALS OF SPECIFIC FUNCTIONS

$$\int x^n \, dx = \frac{1}{n+1} x^{n+1} + C, \, n \neq -1 \qquad \int \frac{1}{x} \, dx = \ln|x| + C$$

$$\int e^x \, dx = e^x + C \qquad\qquad \int a^x \, dx = \frac{a^x}{\ln a} + C$$

$$\int \sin x \, dx = -\cos x + C \qquad \int \cos x \, dx = \sin x + C$$

$$\int \sec^2 x \, dx = \tan x + C \qquad \int \csc^2 x \, dx = -\cot x + C$$

$$\int \sec x \tan x \, dx = \sec x + C \qquad \int \csc x \cot x \, dx = -\csc x + C$$

$$\int \frac{dx}{\sqrt{1-x^2}} = \sin^{-1} x + C \qquad \int \frac{-dx}{\sqrt{1-x^2}} = \cos^{-1} x + C$$

$$\int \frac{dx}{1+x^2} = \tan^{-1} x + C$$

AREA BETWEEN TWO CURVES

$$A = \int_a^b |f(x) - g(x)| \, dx$$

VOLUME OF REVOLUTION ABOUT THE X-AXIS

$$V = \pi \int_a^b [f(x)]^2 \, dx$$

135

Sample examination

Instructions
* Provide complete solutions for full marks.
* Non-programmable, non-graphing calculators are permitted.
* All answers must be in lowest terms.
* Where applicable, round final answers to 3 decimal places.

1. Evaluate each of the following limits.

   (a) $\lim\limits_{x \to -2} \left(4x^3 - 3x^2 + 5x - 9\right)$

   (b) $\lim\limits_{x \to 2} \dfrac{4 - x^2}{2 - x}$

   (c) $\lim\limits_{x \to 3} \dfrac{|x - 3|}{x - 3}$

   (d) $\lim\limits_{x \to \infty} \dfrac{2x^2 - 5x + 1}{4x^2 + 3x - 6}$

   (e) $\lim\limits_{x \to 0} \dfrac{\sqrt{x + 9} - 3}{x}$

   (f) $\lim\limits_{x \to 0} \dfrac{\sin(2x)}{x}$

2. Use first principles to find the derivative of each of the following. State any necessary restrictions.

(a) $y = \dfrac{1}{x-2}$

(b) $y = \sqrt{x+1}$

3. Determine the derivative of each of the following. Simplify fully.

(a) $f(x) = 5 + 7x - 3x^2 - 4x^3$

(b) $h(x) = (2x+1)(x^3 - 8)^4$

(c) $m(x) = \dfrac{4x}{x^2 + 1}$

(d) $y = \sqrt{x^2 - 4x}$

(e) $y = \sin(3x^2)$

(f) $f(x) = e^{\tan x}$

(g) $g(x) = \ln(\cos x) \cdot 2^x$

(h) $\tan^2 y = \sec x$

(i) $x^{-2} + xy = 1$

4. Determine an equation of the tangent to the curve with equation:

$$y = 3x^{\frac{4}{3}} - \frac{3}{2} x^{\frac{2}{3}} \quad \text{at the point where } x = 8.$$

5. The displacement, in metres, of a particle, is given by the equation:

$$s = \sqrt{4t^2 + 1} \quad \text{where } t \text{ represents time, in seconds.}$$

Determine the velocity and acceleration:
   (a) after t seconds.
   (b) after 10 seconds.

6. A 3 m ladder was resting against a vertical wall. It is now sliding downward at a rate of 0.2 m/s. At what rate is the base of the ladder sliding outward, when the top is 1 m from the ground Include a diagram.

7.   Given the function with equation $y = \dfrac{1}{x^2 - 4}$

(a)   Analyse the function under the following headings:
   (i)   local maximum and minimum points;
   (ii)   inflection points;
   (iii)   vertical and horizontal asymptotes;

(b)   Sketch a neat, labelled graph of the function. Include the information from part (a).

8.   The average cost, in $/km, to operate a particular bus is given as

$$C = \dfrac{40}{v} + \dfrac{v}{160}$$ where v is the average speed, in km/h.

Determine the average speed at which the bus should be operated in order to minimize costs. State the minimum operating cost.

9.   A toy tugboat is launched from the side of a pond and travels north at 5 cm/s. At the same moment, a toy yacht begins from a point 800 cm east of the tugboat and travels west at 7 cm/s. How closely do the two boats approach each other?

10.   A searchlight is stationed on the ground and is following a runner who is running at 15 km/h along a straight path that is 100 m from the searchlight at its closest point. What is the rate of change of the rotational angle of the searchlight, when the runner is 800 m from the searchlight?

11.   Determine each of the following integrals.

(a) $\int \left(4x^2 - 5x + 2 \right) dx$ (b) $\int x\left(x^2 - 5 \right) dx$

(c) $\int \sin x \cos x \, dx$ (d) $\int 8x \sin x \, dx$

(e) $\int \dfrac{dx}{x \ln x}$ (f) $\int 6x e^{x^2} \, dx$

138

12. A ball is tossed upward on the planet SELOC, where acceleration due to gravity is 8 m/s². The ball is tossed from a height of 1.5 m at an initial velocity of 12 m/s. When will the ball land?

13. In a chemical reaction, carbon is burned off such that the rate of change of the carbon's mass is proportional to its mass. In 1 hour, 60% of the carbon's mass was burned off.
    (a) What percentage of the carbon was burned off after 2 hours?
    (b) What is the rate of change of the mass of carbon after 4 hours?

14. Determine the area between the curves $y = 2x^2 - 8$ and $y = 4 - x^2$. Include a sketch of the required region, labelling fully.

15. Determine the volume of the solid obtained when $y = 2xe^{2x^3}$ is rotated about the x axis between $x = 0$ and $x = 1$. Include a sketch of the region, labelling fully.

Solutions to exercises

EXERCISE 1

1. (a) 13   (b) 0   (c) -5

    (d) 4   (e) 1   (f) $2\sqrt{11}$

    (g) $\dfrac{1}{2\sqrt{5}}$ (h) $-\dfrac{1}{4}$ (i) 2

    (j) $\dfrac{9}{2}$ (k) 10   (l) $-\dfrac{1}{16}$

    (m) ∞ ∴ does not exist   (n) ∞ ∴ does not exist

    (o) ∞ ∴ does not exist   (p) 0

2.
    (a) $\displaystyle\lim_{x\to 5^+}\frac{|x-5|}{x-5}=1$ $\displaystyle\lim_{x\to 5^-}\frac{|x-5|}{x-5}$ does not exist

    ∴ $\displaystyle\lim_{x\to 5}\frac{|x-5|}{x-5}$ does not exist

    (b) $\displaystyle\lim_{x\to 2^-}f(x)=7$ $\displaystyle\lim_{x\to 2^+}f(x)=4$

    ∴ $\displaystyle\lim_{x\to 2}f(x)$ does not exist   Discontinuous at $x=2$

    (c) $\displaystyle\lim_{x\to 6^+}\sqrt{x-6}=0$ $\displaystyle\lim_{x\to 6^-}\sqrt{x-6}$ is undefined

    ∴ $\displaystyle\lim_{x\to 6}\sqrt{x-6}$ does not exist

    (d) ∞ ∴ does not exist   (e) ∞ ∴ does not exist

EXERCISE 2

1.

(a) $\dfrac{dy}{dx} = \lim\limits_{\Delta x \to 0} \dfrac{\left[3(x+\Delta x)^2 + 2\right] - \left[3x^2 + 2\right]}{\Delta x}$

$= 6x$

(b) $\dfrac{dy}{dx} = \lim\limits_{\Delta x \to 0} \dfrac{\sqrt{(x+\Delta x)-5} - \sqrt{x-5}}{\Delta x}$

$= \dfrac{1}{2\sqrt{x-5}}$

(c) $\dfrac{dy}{dx} = \lim\limits_{\Delta x \to 0} \dfrac{\dfrac{1}{(x+\Delta x)^2 - 1} - \dfrac{1}{x^2-1}}{\Delta x}$

$= \dfrac{-2x}{\left(x^2-1\right)^2}$

2.

$\dfrac{dy}{dx} = \lim\limits_{\Delta x \to 0} \dfrac{\left[6(x+\Delta x)^2 - 9\right] - \left[6x^2 - 9\right]}{\Delta x}$

$= 12x$

$\left.\dfrac{dy}{dx}\right|_{x=2} = 24$

$\therefore y = 24x - 33$

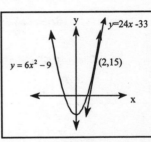

3.

$\dfrac{dy}{dx} = \lim\limits_{\Delta x \to 0} \dfrac{\dfrac{2}{(x+\Delta x)^2} - \dfrac{2}{x^2}}{\Delta x}$

$= \dfrac{-4}{x^3}$

$\left.\dfrac{dy}{dx}\right|_{x=-1} = 4$

$\therefore y = 4x + 6$

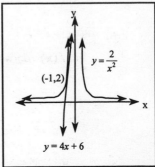

4.

    (a) 14 m/s   (b) 92 m/s

EXERCISE 3

1.
(a) $\dfrac{dy}{dx} = 15x^2 - 8x + 7$ (b) $f'(x) = -16x^{-3} - 12x^{-4}$

(c) $\dfrac{dy}{dx} = \dfrac{9(x-4)}{2\sqrt{x-6}}$ (d) $h'(x) = 6(3x^4 - 5x)^5(12x^3 - 5)$

(e) $\dfrac{dy}{dx} = \dfrac{-2(x^2-1)}{(x^2+1)^2}$ (f) $\dfrac{dy}{dx} = (x+5)^3(15x+23)$

(g) $k'(x) = \dfrac{1}{3}x^2(18x^2 + 7)(2x^3 + x)^{-\frac{2}{3}}$

(h) $\dfrac{dy}{dx} = \dfrac{3(x^2+3x)^2(2x^2+6x+9)}{(2x+3)^4}$

2.
$$\frac{dy}{dx} = \frac{3}{(x-2)^2} - \frac{6}{x^2}$$

$$\frac{d^2y}{dx^2} = \frac{12}{x^3} - \frac{6}{(x-2)^3}$$

$$\frac{d^3y}{dx^3} = \frac{18}{(x-2)^4} - \frac{36}{x^4}$$

3.
(a) $\dfrac{dy}{dx} = \dfrac{-x}{y}$

$\dfrac{dy}{dx}\Big|_{\substack{x=-3 \\ y=5}} = \dfrac{3}{5}$

$\therefore 3x - 5y + 19 = 0$

(b) $\dfrac{dy}{dx} = \dfrac{4y - 3x^2}{y - 4x}$

$\dfrac{dy}{dx}\Big|_{\substack{x=2 \\ y=1}} = \dfrac{8}{7}$

$\therefore 8x - 7y - 9 = 0$

142

EXERCISE 4

1. (a) $v(t) = 12t^2 - 16t + 1$ $a(t) = 24t - 16$

    (b) $v(5) = 221$ m/min $a(5) = 104$ m/min^2

    (c) $t = \dfrac{4 \pm \sqrt{13}}{6}$ min

2. $v(t) = \dfrac{6}{\sqrt{12t - 8}}$

    $a(t) = \dfrac{36}{\sqrt{(12t - 8)^3}}$

    $a(6) = 0.07$ m/s^2

3.

INTERVAL	VELOCITY	ACCELERATION
A to B	positive, increasing	positive
B to C	positive, decreasing	negative
C to D	negative, decreasing	negative
D to E	negative, increasing	positive
E to F	positive, increasing	positive
F to G	positive, decreasing	negative
G to H	zero	zero

4. $P'(t) = -0.048 \times 10^8 (1 + 0.02t)^{-3}$

    $P'(5) = -3\ 606\ 311\, fish/year$

5. $P'(x) = 0.01x - 10$

    $P(100\ 000) = \$48\ 950\ 000$

    $P'(100\ 000) = \$990/item$

6. $\dfrac{dS}{dt} = 8\pi r \dfrac{dr}{dt}$

    $\left. \dfrac{dS}{dt} \right|_{\substack{r=400 \\ \frac{dr}{dt}=2}} = 20\,106.2$ cm^2/min

7.
$$\frac{dz}{dt} = \frac{x}{2}\frac{dx}{dt} + \frac{y}{2}\frac{dy}{dt}$$

$$\left.\frac{dz}{dt}\right|_{t=5} = -25 \text{ m/s}$$

8.
$$\frac{dV}{dt} = \frac{1}{25}\pi h^2 \frac{dh}{dt}$$

$$\left.\frac{dh}{dt}\right|_{\substack{h=2 \\ \frac{dV}{dt}=20}} = 39.8 \text{ m/min}$$

EXERCISE 5

1.

INTERVAL	CONCAVITY
A to B	downward
B to C	upward
C to D	downward
D to E	upward

2. (a) Vertical: $x = 1$ Horizontal: $y = 0$

(b) Vertical: $x = \pm 3$ Horizontal: $y = 1$

3.
$$\frac{dy}{dx} = 3x^2 - 3$$

$$\frac{d^2y}{dx^2} = 6x$$

Domain $= \left\{x \mid x \in \Re\right\}$

Maximum: $(-1,4)$

Minimum: $(1,0)$

Inflection point: $(0,2)$

Concave up: $0 < x < \infty$

Concave down: $-\infty < x < 0$

Intercepts: $(0,2)$, $(-2,0)$, $(1,0)$

Asymptotes: none

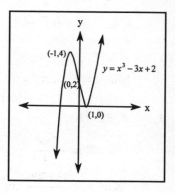

144

4.

$$\frac{dy}{dx} = \frac{-x^2 - 1}{\left(x^2 - 1\right)^2}$$

$$\frac{d^2y}{dx^2} = \frac{2x\left(x^2 + 3\right)}{\left(x^2 - 1\right)^3}$$

Domain $= \left\{x \mid x \neq \pm 1, x \in \Re\right\}$

Maximum / minimum: none

Inflection point: $(0,0)$

Concave up: $-1 < x < 0,\ 1 < x < \infty$

Concave down: $-\infty < x < -1,\ 0 < x < 1$

Horizontal asymptote: $y = 0$

Vertical asymptote: $x = \pm 1$

Intercept: $(0,0)$

EXERCISE 6

1. $C'(x) = 0.000\,25 - 10x^{-2}$

    $x = 200$ pizzas

    Min. $C(200) = \$8.10 /$ pizza

2. $A = \frac{1}{2}b\sqrt{100 - \frac{b^2}{4}}$

    Max. $A = 50$ cm^2 at $b = \sqrt{50}$ cm and $h = \sqrt{50}$ cm

3. $z^2 = \left(15 - 12t\right)^2 + \left(9t\right)^2$

    Min. $z = 9$ km at $t = 0.8$ h

4. $V = 2w^2h$

    $= 3200w - \frac{2w^3}{3}$

    Max. $V = \frac{256\,000}{3}$ cm^3

    at $w = 40$ cm, $h = \frac{80}{3}$ cm, $l = 80$ cm

EXERCISE 6 (cont'd)

5.
$$C = 150(300 - x) + 200\sqrt{x^2 + 2500}$$
Min. $C = \$51\,614.37$ at $x = 56.69$ m,
or 243.31 m down the road from A.

EXERCISE 7

1. (a) $\dfrac{1}{2}$ (b) 0 (c) 0 (d) 1

2. (a) $\dfrac{dy}{dx} = 2\cos(2x)$

   (b) $\dfrac{dy}{dx} = -3\cos^2 x \sin x$

   (c) $\dfrac{dy}{dx} = 3x^2 \sec(x^3)\tan(x^3)$

   (d) $\dfrac{dy}{dx} = 5\sec^2(5x - 3)$

   (e) $\dfrac{dy}{dx} = \dfrac{1}{2}\sin x\left[4\cos x\cos\left(\dfrac{x}{2}\right) - \sin x\sin\left(\dfrac{x}{2}\right)\right]$

   (f) $\dfrac{dy}{dx} = \cos x$

   (g) $\dfrac{dy}{dx} = \dfrac{\cos x}{\sec^2 y}$

   (h) $\dfrac{dy}{dx} = \dfrac{-\sin^3(4y^2)}{16y\cos(4y^2)}$

   (i) $\dfrac{dy}{dx} = \dfrac{-10x}{\sqrt{1 - 25x^2}}$

   (j) $\dfrac{dy}{dx} = \dfrac{2\tan^{-1} x}{1 + x^2}$

3.
$$\dfrac{dy}{dx} = -2\sin(2x)$$

$$\left.\dfrac{dy}{dx}\right|_{x=\frac{5\pi}{6}} = \sqrt{3}$$

$$\therefore 6\sqrt{3} - 6y - 5\sqrt{3}\pi + 15 = 0$$

EXERCISE 7 (cont'd)

4. $\dfrac{dy}{dx} = 2\sin x \cos x - \dfrac{1}{2}$

$\dfrac{d^2 y}{dx^2} = 2\cos 2x$

Minimum: $(\dfrac{\pi}{12}, -0.06)$

Maximum: $(\dfrac{5\pi}{12}, 0.28)$

Inflection point: $(\dfrac{\pi}{4}, 0.11)$

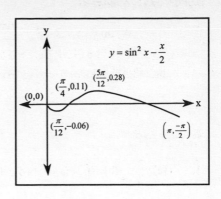

5. $x = 6\sin\theta$

$\dfrac{dx}{dt} = 6\cos\theta\dfrac{d\theta}{dt}$

$\dfrac{d\theta}{dt}\bigg|_{\substack{x=\sqrt{27} \\ \frac{dx}{dt}=0.2}} = 0.067 \text{ rad/s}$

6. $A = \dfrac{1}{2}x\sqrt{10^2 - x^2}$

$\dfrac{dA}{dx} = \dfrac{50 - x^2}{\sqrt{100 - x^2}}$

When $x = \sqrt{50}$ cm,

maximum area $= 25 \text{ cm}^2$

EXERCISE 8

1. (a) $\dfrac{dy}{dx} = 3e^{3x}$

    (b) $\dfrac{dy}{dx} = \dfrac{1}{x - 8}$

    (c) $\dfrac{dy}{dx} = 2e^x(x + 1)$

    (d) $\dfrac{dy}{dx} = e^{x^3}\left(\dfrac{1}{x} + 3x^2\ln x\right)$

    (e) $\dfrac{dy}{dx} = \dfrac{2x}{x^2 - 5}$

    (f) $\dfrac{dy}{dx} = e^{x^2}\left(2x^2 - 2x\ln x + 1 - \dfrac{1}{x}\right)$

    (g) $\dfrac{dy}{dx} = 2e^{2x}\sec^2\left(e^{2x}\right)$

    (h) $\dfrac{dy}{dx} = 2ye^{2x}$

    (i) $\dfrac{dy}{dx} = 4\left(2^{x^4}\right)(\ln 2)x^3$

    (j) $\dfrac{dy}{dx} = \dfrac{\tan x}{\ln 10}$

    (k) $\dfrac{dy}{dx} = \dfrac{e^{5x}\left[5x\ln(5x) - 1\right]}{x\left[\ln(5x)\right]^2}$

    (l) $\dfrac{dx}{dy} = \dfrac{(x - 1)\log_2(x - 1) - (x + 1)\log_2(x + 1)}{\ln 2(x^2 - 1)[\log_2(x - 1)]^2}$

EXERCISE 8 (cont'd)

2. $P = 50\,000e^{\frac{\ln 1.5}{15}t}$

$P\Big|_{t=40} = 147417$ people

3. $D = \{x | x > 0, x \in \Re\}$

$\dfrac{dy}{dx} = \dfrac{2\ln x}{x}$

$\dfrac{d^2y}{dx^2} = \dfrac{2 - 2\ln x}{x^2}$

Minimum: (1,0)

Maximum: none

Inflection point: (e,1)

Vertical asymptote: $x = 0$

Horizontal asymptote: none

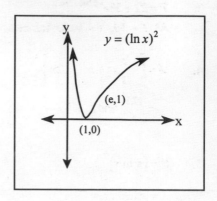

$y = (\ln x)^2$

(e,1)

(1,0)

EXERCISE 9

1. (a) $x^4 - \dfrac{5}{2}x^2 + C$

    (b) $\dfrac{1}{2}e^{2x} - \dfrac{1}{2}e^{-2x} + C$

    (c) $\ln|x| + \dfrac{2}{x} + \dfrac{3}{2x^2} + C$

    (d) $\dfrac{1}{3}\sin x + C$

    (e) $\left[\dfrac{2}{3}\left(x^2 - 5\right)^{\frac{3}{2}}\right]_3^{13} = 1394.8165$

    (f) $x\tan x + \ln|\cos x| + C$

    (g) $e^x(5x - 3) + C$

    (h) $\left[x(\ln x)^2 - 2x\ln x + 2\right]_1^e$
    $= e - 2$

    (i) $\dfrac{2}{3}x(x+2)^{\frac{3}{2}} - \dfrac{4}{15}(x+2)^{\frac{5}{2}} + C$

    (j) $8\ln|x - 5| + 3\ln|x + 1| + C$

    (k) $\left[2\ln|x - 2| - \dfrac{1}{x - 2}\right]_3^4 = 2\ln 2 + \dfrac{1}{2}$

    (l) $\sin^{-1}\left(\dfrac{x}{4}\right) + C$

    (m) $\left[\sin x - \dfrac{1}{3}\sin^3 x\right]_0^\pi = 0$

EXERCISE 10

1. $y = \dfrac{4}{3}\sqrt{x^3} + 6x - 7 - \dfrac{8\sqrt{2}}{3}$

2. $s = -4.9t^2 + 25t + 46$

   At $s = 0$, $t = 6.54$ seconds

3. $T = 22 + 58e^{-0.042286t}$

   At $T = 30$, $t = 46.85$ minutes

4. $A = \left[-\dfrac{2}{3}x^3 + x^2 + 12x \right]_{-2}^{3}$

   $= \dfrac{125}{3}$ units^2

5. $V = \left[\pi \ln x \right]_{2}^{6}$

   $= \pi \ln 3$ units^3

6. (a) $I = -\sqrt{3} \sin 120\pi t + \cos 120\pi t$

   (b) 2 amps

7. $P(x) = \dfrac{1}{3}\sqrt{(x^2 + 25)^3}$

   $P(50) = \$42\ 293.23$

APPENDIX FOUR

Solutions to sample examination

1. (a) $\displaystyle\lim_{x \to -2}\left(4x^3 - 3x^2 + 5x - 9\right) = 4(-2)^3 - 3(-2)^2 + 5(-2) - 9$

$$= -63$$

(b) $\displaystyle\lim_{x \to 2}\frac{4 - x^2}{2 - x} = \lim_{x \to 2}(2 + x)$

$$= 4$$

(c) $\displaystyle\lim_{x \to 3^-}\frac{|x - 3|}{x - 3} = -1 \qquad \lim_{x \to 3^+}\frac{|x - 3|}{x - 3} = 1$

$\therefore \displaystyle\lim_{x \to 3^-}\frac{|x - 3|}{x - 3}$ is undefined

(d) $\displaystyle\lim_{x \to \infty}\frac{2x^2 - 5x + 1}{4x^2 + 3x - 6} \times \frac{\dfrac{1}{x^2}}{\dfrac{1}{x^2}} = \lim_{x \to \infty}\frac{2 - \dfrac{5}{x} + \dfrac{1}{x^2}}{4 + \dfrac{3}{x} - \dfrac{6}{x^2}}$

$$= \frac{2 - 0 + 0}{4 + 0 - 0}$$

$$= \frac{1}{2}$$

(e) $\displaystyle\lim_{x \to 0}\frac{\sqrt{x + 9} - 3}{x} \times \frac{\sqrt{x + 9} + 3}{\sqrt{x + 9} + 3} = \lim_{x \to 0}\frac{x + 9 - 9}{x\left(\sqrt{x + 9} + 3\right)}$

$$= \frac{1}{\sqrt{9} + 3}$$

$$= \frac{1}{6}$$

(f) $\displaystyle\lim_{x\to 0}\frac{\sin(2x)}{x} = \lim_{x\to 0}\frac{2\sin(2x)}{2x}$

$$= 2\times 1$$
$$= 2$$

2. **(a)** $\displaystyle\frac{dy}{dx} = \lim_{\Delta x\to 0}\frac{\dfrac{1}{x+\Delta x-2}-\dfrac{1}{x-2}}{\Delta x}$

$$= \lim_{\Delta x\to 0}\frac{x-2-(x+\Delta x-2)}{\Delta x(x+\Delta x-2)(x-2)}$$

$$= \lim_{\Delta x\to 0}\frac{-1}{(x+\Delta x-2)(x-2)}$$

$$= \frac{-1}{(x-2)^2}, \quad x\neq 2$$

(b) $\displaystyle\frac{dy}{dx} = \lim_{\Delta x\to 0}\frac{\sqrt{x+\Delta x+1}-\sqrt{x+1}}{\Delta x}\times\frac{\sqrt{x+\Delta x+1}+\sqrt{x+1}}{\sqrt{x+\Delta x+1}+\sqrt{x+1}}$

$$= \lim_{\Delta x\to 0}\frac{x+\Delta x+1-(x+1)}{\Delta x\left(\sqrt{x+\Delta x+1}+\sqrt{x+1}\right)}$$

$$= \lim_{\Delta x\to 0}\frac{1}{\left(\sqrt{x+\Delta x+1}+\sqrt{x+1}\right)}$$

$$= \frac{1}{2\sqrt{x+1}}, \quad x>-1$$

3. **(a)** $f'(x) = 7-6x-12x^2$

(b) $\displaystyle\frac{dh}{dx} = 2\left(x^3-8\right)^4+(2x+1)4\left(x^3-8\right)^3\left(3x^2\right)$

$$= \left(x^3-8\right)^3\left[2\left(x^3-8\right)+12x^2(2x+1)\right]$$

$$= 2\left(x^3-8\right)^3\left(13x^3+6x^2-8\right)$$

3. (c) $m'(x) = \dfrac{4(x^2+1) - 4x(2x)}{(x^2+1)^2}$

$\qquad = \dfrac{4 - 4x^2}{(x^2+1)^2}$

(d) $\dfrac{dy}{dx} = \dfrac{1}{2}(x^2 - 4x)^{-\frac{1}{2}}(2x-4)$

$\qquad = \dfrac{x-2}{\sqrt{x^2-4x}}$

(e) $\dfrac{dy}{dx} = 6x\cos(3x^2)$

(f) $f'(x) = e^{\tan x}\sec^2 x$

(g) $g'(x) = \dfrac{1}{\cos x}(-\sin x)2^x + \ln(\cos x)2^x \ln 2$

$\qquad = 2^x(-\tan x - \ln(\cos x)\ln 2)$

(h) $2\tan y \sec^2 y \dfrac{dy}{dx} = \sec x \tan x$

$\qquad \dfrac{dy}{dx} = \dfrac{\sec x \tan x}{2\tan y \sec^2 y}$

(i) $-2x^{-3} + y + x\dfrac{dy}{dx} = 0$

$\qquad \dfrac{dy}{dx} = \dfrac{2x^{-3} - y}{x}$

$\qquad \dfrac{dy}{dx} = 2x^{-4} - \dfrac{y}{x}$

4. $\dfrac{dy}{dx} = 4x^{\frac{1}{3}} - x^{-\frac{1}{3}}$

$\dfrac{dy}{dx}\bigg|_{x=8} = 7.5$

$y\bigg|_{x=8} = 42$

$\dfrac{y-42}{x-8} = 7.5$

$y = 7.5x - 18$

5. (a) $\dfrac{ds}{dt} = \dfrac{1}{2}\left(4t^2 + 1\right)^{-\frac{1}{2}}(8t)$

$\phantom{\dfrac{ds}{dt}} = \dfrac{4t}{\sqrt{4t^2 + 1}}$

$\dfrac{d^2s}{dt^2} = \dfrac{4\sqrt{4t^2+1} - 4t\left(\dfrac{1}{2}\right)\left(4t^2+1\right)^{-\frac{1}{2}}(8t)}{4t^2+1}$

$\phantom{\dfrac{d^2s}{dt^2}} = \dfrac{4\left(4t^2+1\right)^{-\frac{1}{2}}\left(4t^2+1-4t^2\right)}{4t^2+1}$

$\phantom{\dfrac{d^2s}{dt^2}} = \dfrac{4}{\left(4t^2+1\right)^{\frac{3}{2}}}$

(b) $\dfrac{ds}{dt}\bigg|_{t=10} = 1.998 \text{ m/s}$

$\dfrac{d^2s}{dt^2}\bigg|_{t=10} = 0.000498 \text{ m/s}^2$

6. Let t represent the time in seconds.

   Let x represent the distance from the foot of the ladder to the wall.

   Let y represent the distance from the top of the ladder to the ground.

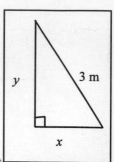

$$x^2 + y^2 = 9 \qquad \text{when } y = 1, \quad x = \sqrt{8}$$

$$2x\frac{dx}{dt} + 2y\frac{dy}{dt} = 0 \qquad \text{given } \frac{dy}{dt} = -0.2$$

$$2\sqrt{8}\frac{dx}{dt} + 2(1)(-0.2) = 0$$

$$\frac{dx}{dt} = 0.0707$$

The base of the ladder is sliding outward at 0.071 m / s.

7.

$$\frac{dy}{dx} = \frac{= 2x}{\left(x^2 - 4\right)^2}$$

$$\frac{d^2y}{dx^2} = \frac{-2\left(x^2 - 4\right)^2 - (-2x)2\left(x^2 - 4\right)(2x)}{\left(x^2 - 4\right)^4}$$

$$= \frac{2\left(x^2 - 4\right)\left[-\left(x^2 - 4\right) + 4x^2\right]}{\left(x^2 - 4\right)^4}$$

$$= \frac{2\left(3x^2 + 4\right)}{\left(x^2 - 4\right)^3}$$

Let $\dfrac{-2x}{\left(x^2 - 4\right)^2} = 0$

$$-2x = 0$$

$$x = 0$$

$$\left.\frac{d^2y}{dx^2}\right|_{x=0} = \frac{2(0+4)}{(0-4)^3}$$

$$= -\frac{1}{8} < 0 \qquad \therefore \text{ max. at } \left(0, -\frac{1}{4}\right)$$

8.
$$\frac{dC}{dv} = -\frac{40}{v^2} + \frac{1}{160}$$

$$\frac{d^2C}{dv^2} = \frac{80}{v^3}$$

Let $-\dfrac{40}{v^2} + \dfrac{1}{160} = 0$

$$v^2 = 6400$$

$$v = 80$$

$$\left.\frac{d^2C}{dv^2}\right|_{v=80} = \frac{1}{6400} > 0$$

$$\therefore \text{ min.}$$

$$C\Big|_{v=80} = 1$$

The minimum operating cost is \$1 / km at an average speed of 80 km / h.

9.   Let x represent the distance between the two boats, in cm.
Let t represent the time, in s.

$$x^2 = (5t)^2 + (800 - 7t)^2$$

$$x^2 = 25t^2 + (800 - 7t)^2$$

$$2x\frac{dx}{dt} = 50t + 2(800 - 7t)(-7)$$

$$2x\frac{dx}{dt} = 148t - 11200$$

For a minimum, let $\dfrac{dx}{dt} = 0$

$$148t - 11200 = 0$$

$$t \doteq 75.7$$

$$x\Big|_{t=75.7} = 465$$

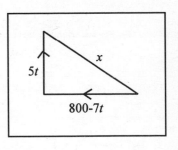

The closest the boats approach each other is 465 cm.

10. Let t represent the time in h.

Let point A be the point on the path closest to the search light.

Let x represent the distance from the runner to point A.

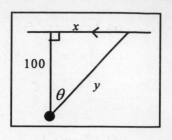

Let y represent the distance from the runner to the light.

Let θ represent the rotational angle of the light between the runner and point A.

$$\tan\theta = \frac{x}{100}$$

$$\sec^2\theta\frac{d\theta}{dt} = \frac{1}{100}\frac{dx}{dt} \qquad \text{Given } \frac{dx}{dt} = -15$$

$$\text{At } y = 800, \quad \sec\theta = \frac{800}{100}$$

$$\sec\theta = 8$$

$$64\frac{d\theta}{dt} = \frac{1}{100}(-15)$$

$$\frac{d\theta}{dt} = -0.0023$$

The rotational angle is changing at -0.0023 rad/h.

11. (a) $\int\left(4x^2 - 5x + 2\right)dx = \frac{4}{3}x^3 - \frac{5}{2}x^2 + 2x + C$

(b) $\int x\left(x^2 - 5\right)dx \qquad \text{Let } u = x^2 - 5 \quad du = 2x\,dx$

$$= \int\frac{1}{2}u\,du$$

$$= \frac{1}{2}\left(x^2 - 5\right) + C$$

11. (c) $\int \sin x \cos x \, dx$ Let $u = \sin x$ $du = \cos x \, dx$

$$= \int u \, du$$

$$= \frac{1}{2} \sin^2 x + C$$

(d) $\int 8x \sin x \, dx$ Let $u = 8x$ $dv = \sin x \, dx$

$$\qquad\qquad\qquad\qquad du = 8dx \quad v = -\cos x$$

$$= -8x + \int 8 \cos x \, dx$$

$$= -8x + 8 \sin x + C$$

(e) $\int \dfrac{dx}{x \ln x}$ Let $u = \ln x$ $du = \dfrac{1}{x} dx$

$$= \int \frac{du}{u}$$

$$= \ln|\ln x| + C$$

(f) $\int 6x e^{x^2} \, dx$ Let $u = x^2$ $du = 2x \, dx$

$$= \int 3e^u \, du$$

$$= 3e^{x^2} + C$$

12. $a = -8$

$v = -8t + 12$

$h = -4t^2 + 12t + 1.5$

Let $-4t^2 + 12t + 1.5 = 0$

     $t = -0.12$ or $t = 3.12$

     $t > 0$ $\therefore t = 3.12$

The ball lands after 3.12 s.

13. (a) Let M represent the remaining fraction of the mass after t

$$M\alpha \frac{dM}{dt}$$

$$\therefore M = M_0 e^{kt}$$

Given $M_0 = 1$

$$M = e^{kt}$$

$$0.4 = e^{k(1)}$$

$$k = \ln 0.4$$

$$M = e^{(\ln 0.4)t}$$

$$M(2) = e^{(\ln 0.4)(2)}$$

$$M = 0.16 = 16\%$$

16% of the mass remained after 2 h.

$\therefore$ 84% was burned off

(b) $\dfrac{dM}{dt} = \ln 0.4 \, e^{(\ln 0.4)t}$

$$\frac{dM}{dt}\bigg|_{t=4} = \ln 0.4 \, e^{(\ln 0.4)(4)}$$

$$= -0.023457$$

The mass is changing at a rate of -2.35% / h.

14. Let $2x^2 - 8 = 4 - x^2$

$$3x^2 = 12$$

$$x = \pm 2$$

$$A = \left| \int_{-2}^{2} (3x^2 - 12) \, dx \right|$$

$$= \left| \left[x^3 - 12x \right]_{-2}^{2} \right| = 32 \text{ units}^2$$

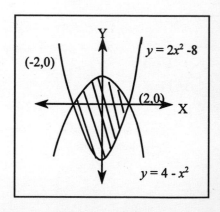

$(-2,0)$ $y = 2x^2 - 8$

$(2,0)$

$y = 4 - x^2$

15. $V = \pi \int_0^1 \left(2xe^{2x^3}\right)^2 dx$

$= \pi \int_0^1 4x^2 e^{4x^3} dx$

$= \pi \left[\frac{1}{3} e^{4x^3}\right]_0^1$

$= \frac{1}{3}\pi\left(e^4 - e^0\right)$

$= \frac{1}{3}\pi\left(e^4 - 1\right) \text{ units}^3$

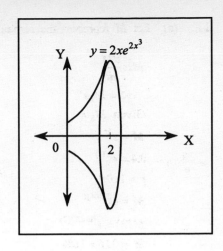

NOTES & UPDATES